PSYCHOLOGY AND MILITARY PROFICIENCY

Psychology
and Military Proficiency

A HISTORY OF THE

APPLIED PSYCHOLOGY PANEL

OF THE NATIONAL DEFENSE

RESEARCH COMMITTEE

By Charles W. Bray

GREENWOOD PRESS, PUBLISHERS
NEW YORK

FOREWORD

MAN-MACHINE is the fighting unit—not man alone, not machine alone. Scientific method applied to making machines suitable to man, and to testing the aptitude of man, selecting and training him to use the machine, has developed so greatly in World War II that instances of this scientific technique should now be set forth for the use of those who succeed us. The aim of this history is to do this so that the scientific method common to all will be apparent and usable in new problems.

One of the greatest intellectual achievements of the war was the application of scientific method to making the material weapon fit the man, and to testing, training, and selecting the man to utilize the particular weapon. The history of the work of the Applied Psychology Panel is written here so that not only psychologists, but any intelligent person in any other field of science or in the armed services, can read it with understanding. The work of our psychologists permeated every field. It made the "man-machine" a fighting unit more effective in the air, on the land, on the sea, and under the sea. I believe that the application of psychology in selecting and training men, and in guiding the design of weapons so they would fit men, did more to help win this war than any other single intellectual activity. A possible exception was the work of the Applied Mathematics Panel of the National Defense Research Committee. Both panels affected many weapons. Other uses of scientific method generally affected only one type of weapon. Application of psychology affected everything from using the atomic bomb to handling cargo.

Long before psychological methods came into use, successful executives from military commanders to top sergeants have had to make judgments on the aptitudes of men to use machines. The very fact that such executives are successful indicates that their judgments are right more often than wrong, but generally the use of scientific methods increases the percentage of correct judgments, and the effect is susceptible of measurement. Even so, this increase is not the principal advantage. Men really successful in making judgments about human beings are few and far between; their methods die with them. Use of scientific methods can be extended to many more men—and does not die with any of them.

Psychologists are somewhat handicapped in their approach to the military or business executive because such executives often prefer to rely on their own insight and personal methods of "skilled appraisal." But in present-day psychology not only experimental techniques but also statistical techniques have advanced so far that they can be and are applied in many cases where skilled appraisal is not enough.

Early in the war it was realized that new sonic apparatus, developed for hunting submarines, was not getting the results that it was physically capable of getting. The trouble lay in selecting and training men to operate the apparatus. A Committee on Selection and Training of Underwater Sound Operators was formed under the guidance of an eminent physicist, Dr. Gaylord P. Harnwell. The membership of the Committee included many psychologists, but it also included physicists, engineers, medical men, etc. It was an outstanding success. More enemy submarines were located for destruction.

The realization arose that many other new weapons could be made more effective if we developed the concept of the "man-machine unit." Dr. Jerome C. Hunsaker, the first Coordinator of Research and Development for the Navy, arranged a meeting of Army officers, Navy officers, and civilian scientists to consider the problem. As a result the National Research Council, in collaboration with the National Defense Research Committee, formed the Committee on Service Personnel—Selection and Training, under the guidance of Mr. John M. Stalnaker as Chairman. Dr. Leonard Carmichael served as the representative of the Council in the preliminary discussions and became a member of the Committee. This new committee was manned by psychologists, but it was to work in close collaboration with physicists, engineers, medical men, psychiatrists, statisticians, and any other scientists who could help. The work of the Committee on Service Personnel was also outstandingly successful.

The work of the Committee on Service Personnel turned out to be needed for nearly all new weapons developed by the nineteen materiel divisions of the National Defense Research Committee. Therefore, as a matter of organizational convenience, the Applied Psychology Panel was set up as a successor to the Committee on Service Personnel. The leaders of this new panel were Dr. Walter S. Hunter, Chief, and Dr. Charles W. Bray, Technical Aide (later Chief). The Panel then worked in closer cooperation with many of the nineteen divisions of the NDRC and collaborated with the Committee on Medical Research in studies for the Army and Navy. The creation of the Applied Psychology Panel met, for psychology, a need which had already been met for mathematics by the Applied Mathematics Panel. Every division of the National Defense Research Committee needed mathematicians and psychologists. If a few men had been assigned to each division, mathematicians and psychologists would have been too thinly spread.

The chapters in the history which follow are written by a psychologist and will be read by psychologists, but they are not written for psychologists alone. They are also written for the intelligent man in any field of activity, be he a military commander, a naval commander, a physicist, an engineer, a chemist, or a business executive. For that reason technical jargon is reduced to a minimum.

You will not read about regression equations or tetrachoric correlations in this history, but other technical ideas will come in. One is the idea of a "test battery." For a simple physical example, if an engineer wants to know the density of a substance, two distinct measurements have to be made: (1) of volume and (2) of mass. The two measurements are completely distinct; they have different and independent "dimensions." They form an elementary test battery.

There are four other technical terms which apply to any test measurement, but apply with peculiar force to psychological tests. These are the terms *reliability, objectivity, independence,* and *validity.*

1. A test is considered reliable if the same observer gets consistent results on repeated measurement of the same individuals or if one random half of a set of measurements gives results consistent with the other random half.

2. A test is considered objective when different observers get consistent results on applying the test independently to the same individuals.

3. A test is considered independent of other tests if it measures different human functions than are measured by the other tests as shown by low correlations between the tests of the battery.

4. A test is valid if it has a high correlation with an acceptable criterion; the test must predict some realistic measurement of performance on a job. Only when it has been proved to do so are psychologists satisfied with it.

Validity has a particular importance because it is sometimes very difficult to find a criterion which we accept as real. If we want a good gunner we can define him as one who makes a certain percentage of hits. If we want a good cargo handler we can accept some score of handling packages in and out of a ship. If we want to select an aviator we might accept as the criterion the passing or failing of the final solo flight. If, as in these cases, the criterion is realistic, the tests which predict it are good tests.

When it comes to some other categories the criterion may be vague. How would we pick a good combat flyer? We know that some men are aces and some soon go down to disaster. How to test them in advance is difficult to know because we do not have very definite measures of what happens while they are in combat. Some studies have been made, and more will be made of the outstanding combat successes and failures, but it is doubtful if we will know for many years how to pick a good combat pilot in advance with the dependability with which we can now pick a man who will pass a training course for pilots.

Likewise with leadership. We know that there are men who are great leaders, but it is difficult to measure leadership on the job. Until it can be measured with reasonable accuracy, the validity of tests for leadership will be unknown.

The concept of validity, with emphasis on a realistic criterion, will appear and reappear in this history. It will be found not only in the chapters dealing with aptitude tests but also in discussions of training and of the design and operation of equipment. The techniques of determination of validity are the chief scientific tools of the psychologist. The manner of their use determines the success of his research.

This history is written for all intelligent men. It is hoped the examples will furnish models for future work. Application of scientific methods by psychologists was a great factor in winning World War II. Future application will help our successors to avoid World War III, or win it if they have the mischance to meet it. In any case, application of psychology to the arts of peace will help our people to a happier life. Happiness requires the adjustment of a man and his capabilities to the world and to other men around him so that he can do something in which he has satisfaction. Psychology can do more to help that adjustment than any other thing I know.

LYBRAND PALMER SMITH
Member, NDRC
Captain U.S.N. (Ret.)

Massachusetts Institute of Technology
Cambridge, Massachusetts
April 2, 1946

PREFACE

IN 1939, at the beginning of World War II, this country's psychologists were engaged in research on many problems of interpretation of mind and behavior. Some were studying how rats learn to run a maze; some were concerned with how we see, hear, or otherwise sense the physical world. Others were doing research on abnormality; some were investigating public opinion; some were interested in the school child, some in the skilled laborer, some in the industrial executive. Almost none were concerned with the special problems of the soldier or sailor.

Beginning in 1939 this country's psychologists were increasingly called upon to use their special knowledge of the human being to promote military proficiency. By the end of 1943, of some forty-five hundred psychologists in this country, over one thousand were engaged in military psychology, in applying their abstract knowledge of human nature to some of the very practical problems of winning a war.

Military psychology was a practical subject. The research of the military psychologist was designed only for practical purposes. It contributed little to an increased understanding of the human being but much to the efficiency of use of the human being. Nevertheless, the work was scientific. It was scientific in the sense that experimental evidence, based on measurement, was sought to prove the value of many practical contributions. Even the hurly-burly confusion of war failed to prevent psychologists from exercising their faith in the practical value of the experimental method. And slowly (as months or even days seem long in wartime) their faith was rewarded. Experimental evaluation provided the basis for a continual improvement of psychological devices and techniques; it prevented mistakes; it proved the practical value of changes in traditional procedures.

Among the organizations which supported research in military psychology was the Applied Psychology Panel of the National Defense Research Committee. The Panel developed from the Committee on Service Personnel—Selection and Training of the National Research Council. Together these groups mobilized some two hundred psychologists for war research. This book is a history of the work of these two hundred psychologists.

Three considerations have governed the writing of this history. The first objective was to provide a record of the more significant contributions for the psychological profession. The second objective was to describe the character of the psychological approach to the problem of military proficiency in order to make its nature clear to fellow scientists and military men. The third objective was to discuss the organizational problems of the war period as a determinant of the war research and as a background to the organization of peacetime military psychology.

In recording the results and describing the character of the psychological approach I have tried to include enough of the evidence and methodology to enable other psychologists to judge the value of the work and the possibilities for future research along the same lines. I have tried to include enough of the evidence and methodology to permit a non-psychological reader to understand that the psychological approach includes a technical search for evidence. I have not attempted to prepare a complete technical account. A complete technical account of the work has already been written in the *Summary Technical Report*[1] of the Applied Psychology Panel. Even though the *Summary Technical Report* is not available to the general reader at this time, it seems unnecessary to duplicate it—particularly because the methods used during the war added little to methodology but only proved again the

---

[1] Dael Wolfle, Editor, *Summary Technical Report of the Applied Psychology Panel. Human Factors in Military Efficiency: I. Aptitude and Classification. II. Training and Equipment.* Washington, D.C., Applied Psychology Panel, NDRC. 1947. This report is not available to the general public.

practical value of existing techniques for the study of human nature.

Chapters 3 to 8 describe the work and the results. The research was carried out by separate little groups of three or four to ten or twelve individuals who worked together on a "project." The typical project was aimed at the improvement of the military proficiency of one or more kinds of personnel. It sought its objective through the development of aptitude tests and classification procedures, through the study of training methods, or through the experimental analysis of the human value of alternative operating procedures or designs for military equipment. Some projects sought to promote military proficiency through one, some through several of these approaches. In the character of their research the projects overlapped, but each project was a unity and each made one or more special contributions. In order to preserve the unity, and yet to avoid repetition, I have treated the projects separately but have made no effort to tell the complete story of every project's work. Rather I have tried to describe a sample, choosing the unique or more important contributions from each project for extended discussion. To a considerable degree, however, this choice of material has been tempered by whether or not the reports of a project are available to the general public. Wherever possible, I have given more extensive treatment to the research which otherwise would be hidden behind the veil of military security.

My third objective has been to describe the wartime need for a coordinating research agency, the organization and policies that were created to meet the need, and the policies which seem to be required if military psychology is to continue its development in peace. Chapters 1, 2, and 9 are devoted to organizational problems. I have treated organization in detail because it concerns the primary problem of any research program, how to obtain and hold first-class research workers. The reader who is uninterested in organizational problems may prefer to skip through these chapters.

In general I have tried to avoid the use of alphabetic abbreviations. In lists and in references, however, I have used OSRD, NDRC, and NRC for the three parent organizations of the Applied Psychology Panel and the Committee on Service Personnel—Selection and Training. The parent organizations were the Office of Scientific Research and Development (OSRD), the National Defense Research Committee (NDRC), and the National Research Council (NRC). The Committee on Service Personnel—Selection and Training has been referred to throughout as the Committee on Service Personnel. I have also abbreviated, as *STR* I and *STR* II, the references to the two volumes of the Panel's *Summary Technical Report*.

The ultimate sources for the factual portions of this book are the reports of the Applied Psychology Panel.[2] Some are cited in the footnotes.

In so far as the reports of the Panel are available to the general public they have been, or will be, abstracted in the *Bibliography of Scientific and Industrial Reports*, Office of Technical Services, U. S. Department of Commerce, Washington, D.C. The Panel reports may be obtained in microfilm or photostat copy from the Office of Technical Services by reference to their abstract number in the bibliography.

My references to the original reports will be found to be curiously incomplete. The reason is that many of the reports cannot be made available to the general public. Military security prevents their general release. Thus, it is permitted to give the source of information on the internal characteristics of the long form of a test, the Personal Inventory, but it is not permitted to give the source for the short form of the same test. The report on the internal characteristics of the short form also contained the test itself. Since the test could not be protected by copyright from improper use through republi-

---

[2] A complete bibliography of all Panel reports given general circulation is contained in Charles W. Bray, *Final Report and Bibliography of the Applied Psychology Panel, NDRC*. OSRD Report 6668. June 30, 1946. Washington, D.C., Applied Psychology Panel, NDRC.

cation, it was protected by the device of declaring its release to be dangerous to military security. This device is quite justified from many points of view but it produces curiosities in a history.

Military security has also required the suppression of some details in reporting experimental results. Thus, it is permitted to report relative but not absolute improvements in performance in the control of gunfire. Despite the unsatisfactory nature of statements of percentages without specification of the absolute values, the latter could not be given without furnishing possible assistance to hypothetical future enemies.

In a few instances reference has been made to the work of other groups in the Office of Scientific Research and Development. The work of the whole organization has been described in J. P. Baxter's *Scientists Against Time*. Under the titles *New Weapons for Air Warfare* and *Applied Physics: Metallurgy; Electronics; Optics*, there will soon appear descriptions of the work of particular psychological groups other than those of the Applied Psychology Panel but nevertheless in the National Defense Research Committee. These three volumes are a part of a series entitled *Science in World War II* from the Little, Brown Publishing Co., Boston, Mass.

The direct sources of most of this book are not available to the general public. First may be mentioned the many informal discussions of military psychology which I have enjoyed with various individuals on the Applied Psychology Panel and with the Panel liaison officers. In particular some of the ideas of Leonard Carmichael, Walter Hunter, John Kennedy, and Dael Wolfle will appear in these pages. The guidance and help of Capt. P. E. McDowell, USN, Capt. (now Prof.) L. P. Smith, USN (Ret.), and Maj. H. E. Clements, AUS, must also have found expression in Panel research. Captain Smith's broad philosophy is exemplified in the Foreword of this book.

My second direct source is the *Summary Technical Report* of the Applied Psychology Panel. The authors of the two volumes of the *Summary Technical Report* were W. C. Biel,

C. W. Bray, B. J. Covner, Norman Frederiksen, W. E. Kappauf, J. L. Kennedy, D. B. Lindsley, and Dael Wolfle. These authors will find that I have borrowed heavily from them, paraphrasing and quoting sentences, paragraphs, and even whole sections so frequently that detailed references and quotation marks are inadvisable. This departure from usual practices may possibly be pardoned in an official history of a cooperative program.

Walter Hunter, John Kennedy, A. T. Poffenberger, Dael Wolfle, and Helen Wolfle read a draft of the manuscript and offered many suggestions and corrections. In particular, Dael Wolfle provided many ideas and helped in organization and expression, while Helen Wolfle smoothed the style. The Army and the contractors of the Applied Psychology Panel supplied the original photographs reproduced in some of the figures. My indebtedness to these individuals and organizations is here acknowledged.

If credit is due for the accomplishments of the Applied Psychology Panel and the Committee on Service Personnel, that credit belongs to the psychologists who conducted the actual research. The members and staff of the Panel and the Committee here record their gratitude to these psychologists as individuals.

CHARLES W. BRAY
Chief, Applied Psychology Panel

Princeton, N. J.
March 9, 1947

CONTENTS

PSYCHOLOGY AND MILITARY PROFICIENCY

CHAPTER 1

THE FORMATION OF THE APPLIED
PSYCHOLOGY PANEL

WHAT are the major problems which the Army and Navy must solve if sound strategy and tactics are to be effective in war? The Army's General Staff and the Navy's Bureaus have been organized to deal with such problems and the titles given to the General Staff divisions, for instance, indicate their nature: Personnel, Intelligence, Training, Operations, Materiel, Supply. A fair share of the work of the General Staff concerns the effective use of manpower.

Manpower and its use are perennial problems. The methods of warfare change; new devices and new explosives are developed, used, and become obsolete. The officers and enlisted men remain. Their intelligence and skill, their preparation for war, their willingness to fight are basic in combat success.

During World War II the armed services turned for help to the science whose research concerns the behavior of the human being. A rapid development of military psychology resulted. Military psychology is the application of psychology to the development of military aptitude tests and classification procedures; to the training of men; to the design of military equipment for human use; to the simplification of military operating procedures; to the problems of the abnormal soldier or sailor; and to psychological warfare, morale, and human attitudes. From military psychology came methods and devices of proven value.

Among the many organizations which contributed to the use of military psychology in World War II was the Applied Psychology Panel of the National Defense Research Committee. The Applied Psychology Panel was created for the purpose of stimulating and correlating the war work of civilian psychologists on classification, training, and the design and operation of equipment. This book describes the history of the Panel—the need for its creation, the administration of its research program, and some of the results of its work.

In 1939 military psychology barely existed. A few psychological tests left over from World War I and the period just following that war were sometimes used at local installations. In 1940 the Adjutant General of the Army set up a Personnel Procedures Section. In 1941 the Army Air Surgeon began the development of a Psychological Branch. Early in 1942 an Aviation Psychology Section was organized by the Navy's Bureau of Medicine and Surgery. Psychologists were recruited in large numbers to these units.

The basic problem of the psychologist in uniform was to classify personnel. Classification is a part of the process by which each man is assigned to duty. When it is well done, each man's peculiar aptitudes, experience and interests are determined so that his assignment may be in accord with his characteristics. In ordinary life the typical process of job assignment and reassignment is empirical rather than technical. Men choose, and are hired for, jobs without a clear understanding of the qualities required. After some time they turn out to be successful or unsuccessful. Trial and error, sometimes pure blind chance, are determining. The classification of men is an attempt to short-circuit this empirical process. All possible information about jobs and men is assembled. Each man can then be assigned to a job according to some system. The psychologist enters this process because he has special techniques to measure or determine relevant aptitudes, experiences and interests and because he can evaluate classification and assignment procedures by experiment.

The methods and knowledge of the psychologist are also useful in training men. The principles of human learning are basic to effective training. The effects of training must be measured to determine the success of any particular application of the principles. In education generally, and in many industries, training is routinely subjected to psychological inquiry. In 1942 the Bureau of Naval Personnel began the development of a research section devoted not only to classification but to training as well.

The mentally sick individual has always plagued the Army and Navy. Psychological warfare, morale, and human attitudes are other fields to which increasing attention is being paid. Clinical psychologists were brought into the services to study the first of these problems; social psychologists were employed to study the second.

In modern war men are trained not only in discipline and the elementary skills of direct combat but also to maintain and operate complex equipment. In World War II, the "physicist's war," military devices were radically revised. Their complexity increased. Thus even the regular officer and enlisted man, as well as the reserve officer and recruit, had to use unfamiliar equipment. And in the early days of the war the regulars were swamped in the process of mass mobilization. Under the circumstances it was inevitable that many new devices were misused. Soldiers and sailors were unable to realize all the potentialities of the new materiel in combat.

The first reaction to the misuse of new equipment was to blame the quality of our personnel and our training methods. High-priority equipment was given a priority on manpower of above-average ability. Special attention was paid to the training of the men who were assigned to duty with the new devices. Psychologists were called upon to study the selection and training of various kinds of specialists.

In the course of research on selection and training the psychologists began to suggest that some of the new equipment was so designed that it made unnecessarily heavy demands on personnel, or that the standard procedures of operation were unnecessarily complex. The psychologists suggested ways of improving equipment and simplifying operating procedures.

The design and operation of equipment is a subject which is not ordinarily associated with psychology. An example may help to define the relation. If one wishes to determine the distance, or range, to a target by optical methods, several kinds of rangefinder can be used. The stereoscopic rangefinder is one type of instrument which permits quite accurate distance measurements. Coincidence rangefinders, using a completely different principle of human vision, are a second type. Which has greater value in the military situation? Obviously the answer depends on optical and mechanical considerations. The excellence and cost of an instrument as a physical device is basic. In a practical case, however, physical considerations may actually be secondary in the choice. The acuity of the human eye in stereoscopic discrimination may be greater than in coincidence discrimination, or it may be less; the two kinds of acuity vary independently of one another as conditions change. The precision of range determinations by either instrument depends on a number of specific design and operational factors which relate the device to human characteristics. In addition, a stereoscopic device requires prolonged training of the operator and only a small proportion of all people ever become proficient with it. The training of the nonproficient is wasted. So stereoscopic operators must be selected. A more efficient device might reduce the need for selection and training. Thus the military worth of any equipment can be assessed only when one knows the answers to several questions about human beings.

The military worth of alternative devices and operating procedures can be evaluated far more adequately if the associated psychological questions are answered quantitatively. Just how accurate is the average human being with each method? How much accuracy is gained by this or that or the other change in design? Does a given change call for a change of training methods? How many months are required for training? What proportion of all men reach an acceptable standard of accuracy in these months? What aptitude test scores must a man have so that the odds favor his reaching this standard after training? What proportion of errors will be made in selecting trainees? The psychologist is prepared to give quantitative answers to these questions about man in relation to the machine, or to secure the answers through research if necessary.

Two civilian war research agencies were particularly interested in the psychological approach to equipment problems. These were the National Defense Research Committee and the National Research Council. The former was a part of an emergency government office, the Office of Scientific Research and Development, which planned, supervised, and coordinated civilian scientific research on military problems. On recommendation of the National Defense Research Committee, this government office entered into research contracts with university and industrial laboratories for the development of new materiel.

The National Research Council was a quasi-governmental agency. It had been established in the first world war under federal charter to advise and carry on research at the request of the government. It was organized under the National Academy of Sciences. In World War II, the National Research Council carried on many research and advisory programs at the request of the War and Navy Departments, the Office of Scientific Research and Development, and other government agencies.

These two civilian scientific agencies, the National Defense Research Committee and the National Research Council, made increasing use of psychologists in connection with problems of materiel and operations. Particular attention was given to the human factor in the operation of stereoscopic heightfinders;[1] to the use of the human eye at night,[2] and to human hearing considered as an integral part of telephone

---

[1] By Division 7, NDRC.
[2] By Division 16, NDRC, the Committee on Medical Research, OSRD, and the medical committees of the National Research Council. The early American research on night vision was greatly stimulated by the work of the British Flying Personnel Research Committee.

and radio communications.[3] In these programs the point of view arose that while it is desirable to fit the man to the machine it is equally desirable to fit the machine to the man. Classification, training, and equipment came to be seen as interrelated aspects of a single military problem—how to get the most effective results from the man-machine combination.

The widening interest in military psychology caused a mushroom growth of psychological research. Within the Army, the Navy, the National Defense Research Committee, the National Research Council, and other agencies, psychologists were put to work. In some instances, notably in the Army and Navy, a single group was large enough to constitute a strong effective organization. In many cases, however, the psychologists were scattered over many non-psychological agencies. The effects of the mushroom growth were failures of liaison and coordination, and limitations on the kinds of research that could be undertaken.

In the fall of 1941 the Office of Scientific Research and Development requested the National Research Council to survey research on night vision. While the survey began with night vision, information was also collected as a by-product on any psychological research organization concerned with classification, training, or equipment.[4]

Although the survey was incomplete, it revealed an extraordinary variety of agencies with a direct interest in supporting psychological research. Over fifty separate groups were counted; these operated under military, scientific, engineering, and medical auspices.

The separate groups were incompletely aware of one an-

---

[3] By Division 17, NDRC.

[4] No formal report of the survey was made. It was conducted by the Committee on Human Aspects of Observational Procedures, NRC: L. Carmichael, Chairman; P. W. Bard and S. W. Fernberger, Members; C. W. Bray, Research Investigator. The results of the survey are incompletely described in a letter, C. W. Bray to L. Carmichael, February 2, 1942.

other's work. In part this reflected the rapid growth of research; in part it resulted from the need for military secrecy.

In each group there were limitations on research due to the limited directives and interests of those in charge. In the armed services the emphasis was on classification and aptitude testing; training was a secondary interest. In the civilian groups there was greater freedom to make whatever seemed to be the most appropriate approach to the man-machine problem. But each civilian group was limited, as the service groups were not, to research on some single kind of military personnel.

Despite the secrecy and the limitations, despite the variety of auspices under which the psychologists worked, there was a striking similarity in the research approach of each successful group. The hallmark of a satisfactory psychological program was the development of an effective criterion measure against which to check the validity of alternative tests, training methods, or equipment designs.

In a number of instances psychological research was under the direction of non-psychologists. The effect was to minimize the psychological approach and, in some cases, to lead psychologists out of psychology altogether. In other cases the result was that industrial or test psychologists worked on problems which required the techniques of the laboratory man, and vice versa. No effective organization existed to classify the psychologists themselves.

A need for another kind of coordination appeared in comparing the research of the civilian groups who were concerned with the selection of some particular kind of personnel. Among such groups there was a heavy reliance on tests of rather general types of aptitudes. For instance, the tests recommended for service use as a result of research on the selection of aircraft pilots, stereoscopic heightfinder operators, and underwater sound (antisubmarine gear) operators were these:

Aircraft Pilots	Heightfinder Operators	Sound Operators
Otis SA Test	Army General Classification Test	Otis SA Test
Bennett Mechanical Comprehension Test	Army Mechanical Aptitude Test	Bennett Mechanical Comprehension Test
Biographical Inventory	Tests of Visual Capacity	Tests of Auditory Capacity
Psychomotor Tests		

The capacities measured in the first two items of each list were closely similar and included verbal, arithmetical, and mechanical aptitudes. These are rather general capacities which were needed by every branch of the Army and Navy. Yet the general tests recommended by the three groups were not the same tests. Furthermore, there were not enough men of high general capacity to meet the needs of every branch of the Army and Navy. Hence the military problem was to obtain precise information on the general aptitude requirements of all specialties, to relate the general requirements to possible special requirements, and to reconcile the needs of all service branches with the available supply of men of high general aptitudes. That is to say, the military problem was one of classification of all men, not the selection of a few. It was unlikely that such a problem could be solved by many isolated research groups. Individual research groups with a loyalty to a given specialty inevitably breed competition for the generally good men. Individual research groups may fail to see the overall problem and the desirable simplification which comes when adequate tests of special ability are available.

Despite the many limitations on the research of the early period it had given results of practical value in the solution of military problems. Better classification simplified training and produced more effective use of equipment. Better training reduced the need for classification and for simpler materiel. Better equipment relieved the strain on classification and training.

Thus the suggestion arose that a strong centralized psychological research organization should be created. Such an organization was not necessary to replace existing groups, whose very growth attested to their value; but it was needed to supplement existing groups and to help carry on liaison between them. There was no logical place for a centralized agency in either service, but the civilian war research agencies were well adapted by their structure for the function of coordination. In March 1942 the suggestion was made to the Office of Scientific Research and Development that it create a general psychological group.[5]

In response the Office of Scientific Research and Development pointed out that psychological research on military problems would be dependent for practical effectiveness on the backing of highly placed officers in one or both services. Military psychology was resulting in proposals for fairly radical modifications of traditional personnel procedures, and only an officer at a high echelon could break through the halo of tradition and secure effective use of research results. Essentially the Office of Scientific Research and Development asked: *Who* is this officer?

In the Navy two such men were known. They were Comdr. (later Capt.) P. E. McDowell, USN, of the Office of the Commander in Chief, U.S. Fleet, and Capt. L. P. Smith, USN, Assistant Coordinator of Research and Development, Office of the Secretary of the Navy Department, and Navy member of the National Defense Research Committee. Each of these men was familiar with the civilian psychological research of the period. Each had a passionate belief in the practical military value of psychological research. Each perceived the need for improved liaison and coordination. Each was in a position to help secure effective use of research results.

In addition Commander McDowell was concerned with two general problems on which the assistance of civilian psychologists was then desired. The first was to revise the apti-

5 Letters, L. Carmichael to V. Bush, March 14, 1942, and May 4, 1942.

tude testing program of the Bureau of Naval Personnel. The tests in use by that Bureau dated from the period just following the first world war. The second was to improve the quality of antiaircraft gunnery. The personnel factor in antiaircraft gunnery had not yet been subjected to a general psychological study. A new civilian research agency might assist in both problems.

From the interests of Commander McDowell and Captain Smith there resulted a conference on June 6, 1942, to discuss the need for a new psychological agency. The War and Navy Departments, the Office of Scientific Research and Development, the National Defense Research Committee and the National Research Council were represented. Following the conference the Navy Department, with the concurrence of the War Department, requested the National Defense Research Committee to establish a general psychological research organization. The request was accepted and a Committee on Service Personnel was set up in the National Research Council under a government contract recommended by the National Defense Research Committee. The work of the Committee began on June 20, 1942, and developed rapidly. The overall picture soon came to be one of continuous expanding demand by the Army and Navy for psychological research under Committee auspices.

The growth of the Committee led to a need for simplification of its own organizational structure. On October 7, 1943, the Committee was dissolved and reconstituted as the Applied Psychology Panel of the National Defense Research Committee. This change increased the efficiency of the group through a simplified organization and an increased staff. The Applied Psychology Panel continued the policies and program of the Committee and all the members and staff of the Committee served on the Panel. Hence the work of the two groups can be considered as a single continuous activity. To simplify the presentation, the term Applied Psychology Panel will include the Committee on Service Personnel in this history of their work, except in those instances in which the administrative details require specific reference to the Committee.

The research of the Applied Psychology Panel concerned those aspects of military psychology which were related to the fundamental purpose of the National Defense Research Committee—the development of instruments of war. The relevant psychological subjects were classification, training, and equipment. Although the clinical and social approaches to military psychology are an integral part of any complete consideration of the human being conceived as an instrument of war, they were not believed to be fit subjects of inquiry for a materiel agency.

Within the general fields of the Panel's research some twenty major research programs were undertaken. These are classified according to the major field of research interest (and without reference to chronology) in Figure 1. At the upper left appears a series of studies concerned primarily with classification. In the upper center are studies which concentrated on training. Equipment studies are listed at the right. As one passes to the bottom of the figure, there appear coordinated studies of several of the major problems of military psychology; in these, classification and training, or training and equipment, were studied simultaneously. Near the bottom are two projects which maintained a coordinated approach to all three subjects. At the bottom is a topic common to nearly all studies of classification, training, and equipment—the achievement and proficiency tests, which are the psychologist's criteria, or measuring sticks of the validity of his methods. Toward the end of the war the achievement and proficiency tests began to receive attention as such.

Figure 1 shows the emphasis on classification and training which characterized the research of the Panel as a whole. This emphasis resulted from the original interests and directives to the Panel and from the strains of mass mobilization. Nevertheless the original emphasis shifted as the Panel developed its own peculiar field and as the services developed their own research organizations. The emphasis on equipment and on achievement and proficiency testing grew steadily. The shift in emphasis is graphically illustrated in Figure 2 which shows the character of the Panel's reports and the changes in

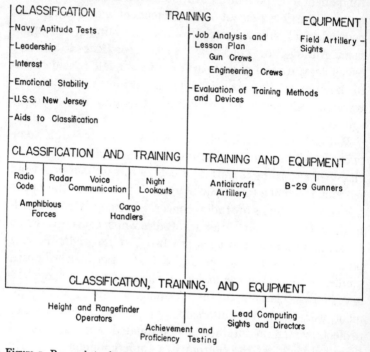

Figure 1. Research topics of the Applied Psychology Panel, June 20, 1942, to October 31, 1945.

character with time. The figure shows, for each six-month period from June 1, 1942, to December 1, 1945, the total number of reports and the numbers in each of the major categories of Panel interest. Many arbitrary decisions were required to classify the reports for presentation in this form, but the graph is accurate in trend. It describes a total of 513 separate documents, including all Panel reports which were given general circulation.

Figure 2 clearly shows the original emphasis on classification and training. In 1943 classification and training received about equal attention; in 1944 training dominated the scene. In the fall of 1944, as the end of the German war was expected and service programs on classification and training became more effective, the Panel deliberately reduced its research on classification and training.

The figure shows a slow but steady increase in Panel research on design and operation of equipment and on achievement and proficiency tests. General reports, a heading which includes final reports on projects, general recommendations and theoretical contributions, were relatively rare until the final period. Then, of course, a considerable number of final reports were prepared.

The overall "curve of production," or total number of reports, is also shown in Figure 2, and it deserves special con-

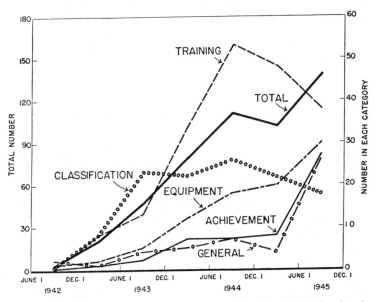

Figure 2. The total number of reports of the Applied Psychology Panel and the number in each of several categories during successive six-month periods from June, 1942, to December, 1945. A total of 513 reports, training manuals, pamphlets, sets of phonograph records, etc. are represented.

sideration. The initial rise in total number of reports was slow. The slowness reflected the time required to establish research projects and the gradual increase in productivity of each project with time. The curve thus illustrates inertia in the mobilization of scientific effort. The Panel could consider the establishment of only a few projects at a time, being limited chiefly by the shortage of top-level personnel to direct the projects. After a project was started, it required a period of several months to a year to assemble a full staff, familiarize the staff members with military problems, develop a laboratory, and turn out effective results.

It should be indicated in this connection that a further delay, varying in length from a few months to a year, necessarily intervened between the completion of research and general military use of the research conclusions. This was true despite the fact that results were often in general use before a formal report of results appeared. When the delay in application of results is added to that shown in Figure 2 it is clear that a psychological program should be organized well in advance of an emergency. If it is organized after the emergency has arrived, much of its possible effectiveness is lost.

If psychological research is already in progress, new research can be completed and results applied with much less delay than when the work must start from scratch. Panel experience in the later days of the war was that new problems could be solved and results applied within a few months by projects already organized. At best, however, neither research nor application can be hurried too much. Needs must be anticipated.

A decline in report production by the Panel began in the late fall of 1944. The decline reflected the termination of a number of projects in anticipation of the end of the German phase of the war. At the time of the German surrender in the spring of 1945, a further reduction was planned. Even had the Japanese continued the war beyond the summer of 1945, the Panel would have had only a few projects in operation. The reduction in the Panel's activities was made possible by the approaching end of the war and by the continued growth of psychological research in service units. The final rise in the curve for total number of reports merely reflects the simultaneous and sudden termination of all activities.

CHAPTER 2

THE ADMINISTRATION OF THE RESEARCH PROGRAM

THE research of the Applied Psychology Panel and its predecessor, the Committee on Service Personnel, was conducted under a complex administrative arrangement.[1] The arrangement was particularly complex in the case of the Committee on Service Personnel; it was less complex in the case of the Applied Psychology Panel. Under either arrangement the personnel, purposes, activities, and general policies were the same. The personnel, the administrative patterns, and the policies of the two organizations may be described in relation to the activities which implemented their common purposes and policies.

PERSONNEL

The Committee on Service Personnel was conceived from the start as a "working" committee, i.e. its members were to be active in research and not merely advisory. Several advisory committees on psychology were already in existence.[2] The need was for a compact working group whose members could be free to undertake direct supervision of a series of coordinated research programs in the fields of classification, training, and equipment. It was therefore desirable that the new Committee be composed of relatively young, active research men from the fields of psychometrics, industrial psychology, and laboratory psychology. The members should be experienced in research administration and familiar with mili-

[1] This chapter is based on Charles W. Bray, *Summary of Activities on Project N-100, The Committee on Service Personnel—Selection and Training, The Committee on Applied Psychology and the War. Final Report on Contract OEMsr 614.* OSRD Report No. 6573. October 31, 1945. National Academy of Sciences. Washington, D.C., The Applied Psychology Panel, NDRC.

[2] e.g. The NRC Emergency Committee in Psychology and the NRC Committee on Military Psychology Advisory to the Adjutant General.

tary needs. The members could receive no pay, but it was necessary that they be free to donate a considerable fraction of their time to the work of the Committee.[3]

The effort to meet these requirements resulted in the civilian membership of the Committee which, with pre-Committee status, is listed as follows:

J. M. Stalnaker, Chairman of the Committee. Associate Secretary of the College Entrance Examination Board and consultant to the Bureau of Naval Personnel.

G. K. Bennett. Director of the Test Division, Psychological Corporation, and consultant in numerous war industries.

Leonard Carmichael, ex-officio member as Chairman of the Division of Anthropology and Psychology, NRC, and Contractor's Technical Representative for the contract covering the work of the Committee. President of Tufts College, active in research for Division 7, NDRC, and Director of the National Roster of Scientific and Specialized Personnel.

C. H. Graham. Associate Professor of Psychology, Brown University, active in research for Division 7, NDRC, and consultant to Division 16, NDRC.

M. S. Viteles. Professor of Psychology, University of Pennsylvania, Chairman of the NRC Committee on Selection and Training of Aircraft Pilots, consultant to Division 6, NDRC, and to numerous industrial concerns and government agencies.

In addition to these men, who formed the working nucleus and later organized as the Subcommittee on Procedures (the executive subcommittee), three service representatives were included in the membership of the main Committee. These were:

[3] The choice of members was automatically limited to the eastern part of the United States by the requirement that the members be active yet unpaid. By count from their travel vouchers the volunteer members of the Committee were on travel status from one-third to one-half of the days in the three months following the formation of the Committee.

W. V. Bingham, Chief Psychologist, Office of the Adjutant General, War Department.

Comdr. P. E. McDowell, USN, Readiness Division, Office of the Commander-in-Chief, U.S. Navy.

Capt. F. U. Lake, USN, Training Division, Bureau of Naval Personnel, Navy Department.[4]

The full-time, professional staff of the Committee included:

C. W. Bray, Executive Secretary of the Committee and Technical Aide in the Office of the Chairman, NDRC, for the business of the Committee. Associate Professor of Psychology, Princeton University, Research Investigator for the NRC Committee on Human Aspects of Observational Procedures, and consultant to Division 6, NDRC.

J. L. Kennedy, Assistant to the Executive Secretary of the Committee. Assistant Professor of Psychology, Tufts College, and active in research for Division 7, NDRC.

ORGANIZATION

The primary work of the Committee was to initiate, supervise, and assist research projects in the fields of classification, training, and equipment. These projects were established in two ways: (1) as direct Committee projects conducted by the Committee's own employees, or (2) as projects under contracts between the National Defense Research Committee and private institutions or laboratories. In the second case, the Committee on Service Personnel formulated a program and recommended the contract to the National Defense Research Committee. When the recommendation was accepted, the Committee on Service Personnel followed the work of the contractor, assisted him, and secured use of the results by carrying on liaison and advisory activities with the Army and Navy. These basic relationships and the agencies through

---

[4] Captain Lake was replaced by the following, who served in turn as representatives of the Bureau of Naval Personnel: Lt. Comdr. H. J. Pohl, USN, Lt. Comdr. R. A. Sentman, USN, and Capt. W. E. Moore, USN.

which they were implemented are shown in the organization chart of the Committee in Figure 3.[5]

As Figure 3 indicates in its central column, the basic relations were those of the fundamental contract between the

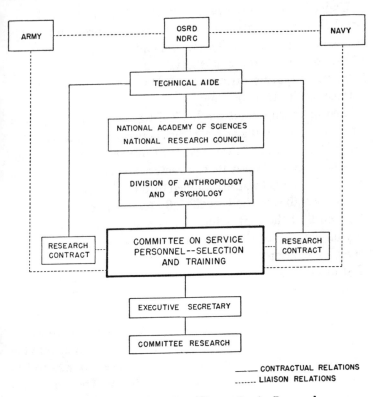

Figure 3. Organization chart of the Committee on Service Personnel.

Government (OSRD–NDRC) and the National Academy of Sciences, National Research Council. From the point of view of the Government the scientific activity under this contract was the direct responsibility of a Technical Aide on the staff

[5] The form of organization of the Committee was patterned on that of the Committee on War Metallurgy of the National Research Council. The pattern was adopted at the suggestion of the Office of Scientific Research and Development.

of the National Defense Research Committee. From the point of view of the National Research Council as contractor the responsibility was in the hands of the Committee on Service Personnel.

Under its Executive Secretary the Committee employed psychologists to carry on direct research programs. This mechanism was generally utilized only in the preliminary phases of a program or in special cases of minor temporary activities requiring little research but considerable advisory service to the Army or Navy.

The more important mechanism was the formal research project organized under a contract between the National Defense Research Committee and some private institution. Contact between these private contractors and the Committee on Service Personnel was maintained through the Technical Aide mentioned above, who was officially responsible for the scientific work of the contractor. It was also maintained through liaison activity of the Committee members themselves. The complex relations shown in Figure 3 were somewhat simplified in practice by the fact that the Technical Aide and the Executive Secretary of the Committee were actually one individual (the present author) working on two separate jobs.

Complex as these relations were, they were complicated still further by the relation between the civilian group and the services. The whole purpose of civilian research was to improve the efficiency of the Army and Navy. The closest cooperation between the Committee and the services was required, but independence of scientific research from military domination was desirable. The liaison relations, the channels through which information flowed from the civilians to the military and from the military to the civilians, were basic to the effectiveness of the Committee. The problems of liaison will be separately discussed below.

The organization of the Committee on Service Personnel was complex. It is fair to say that it was understood by the Committee and its parent organizations but by few others.

Many hours were spent in attempting to explain the relationships shown in Figure 3 to liaison officers and research staffs. The similarity of the names, National Defense Research Committee and National Research Council, led to minor misunderstanding and to delay in correspondence.

Nevertheless, the organization provided an extraordinary degree of flexibility in setting up research under government auspices. The flexibility was such that almost any kind of subordinate organization for the actual conduct of research or liaison was possible. Many were tried in the months immediately following the formation of the Committee, for the work of the Committee grew with unforeseen rapidity, but with a sufficiently sound basis nevertheless. The original organization and the even more complicated relations which followed may have been confusing, but they failed to stop the flight of what, at the time, was essentially a trial balloon.

The research grew. Over the summer of 1942 the Committee surveyed the military scene and carried on a continuous series of active advisory relations with an increasing number of liaison officers. In September the Committee established two research projects under OSRD contracts. Through the fall of the year it mobilized three additional contract groups. In the winter of 1943 two more projects were created, and a third, a cooperative program with other agencies, was begun and completed. Over the summer four more were organized. Throughout the whole period the Committee carried on a continuous series of programs through its own expanding staff. It established and maintained close liaison with a number of NDRC divisions. It rejected a number of proposals for research which appeared to be impractical or not appropriate for Committee support. Growth was limited chiefly by the rate at which competent research personnel could be located and persuaded to leave their academic laboratories for the somewhat more rigorous life of traveling men or for the less satisfactory (to the scientist) research possible under wartime conditions in military establishments.

Growth increased the organizational complexity. Two new subcommittees were formed. It was intended that each should guide a project.[6]

Growth created a need for a larger full-time staff of general administrators. Nevertheless the latter remained at its original size of two. Continual additions to the staff were made but all proved to be temporary. Each added staff member was soon sacrificed to the needs of some additional specific research program and was placed on a contract.

Under the circumstances discussion of reorganization began within a few months of the first meeting of the Committee on Service Personnel. Over the winter of 1943 the discussion became more concrete and in July 1943 it resulted in a conference between Committee representatives and other interested divisions of the National Defense Research Committee. The conference led to the organization of the Applied Psychology Panel of the National Defense Research Committee. The functions of the Panel were ultimately defined as the continuation of the work of the Committee on Service Personnel. The change to a panel consisted entirely of a simplification of organizational relations and the addition of new personnel.

The new organization is described in the simple chart of Figure 4. The Panel became one of several similar divisions and panels operating directly under the National Defense Research Committee. The Chief of the Panel, with the help of his full-time Technical Aides and the advice of the volunteer Panel members, was responsible for the initiation of research contracts, delivery of results, and liaison with the services.

The Chief of the Panel was W. S. Hunter, Chairman of the Department of Psychology at Brown University and a man

---

[6] The Subcommittees and their membership were: Subcommittee on the Selection and Training of Oscilloscope Operators; G. K. Bennett, Chairman, Comdr. (later Capt.) D. C. Beard, USN, G. A. Fry, H. K. Hartline, Don Lewis, H. W. Nissen, and L. L. Thurstone. Subcommittee on Voice Communication; A. T. Poffenberger, Chairman, G. K. Bennett, Don Lewis, S. S. Stevens.

active at the time in a number of war projects. The Panel membership included the entire non-service group from the Committee on Service Personnel. The Executive Secretary of the Committee and his Assistant became Technical Aides to the Panel Chief. Dael Wolfle, Associate Professor of Psychology at the University of Chicago and active in the train-

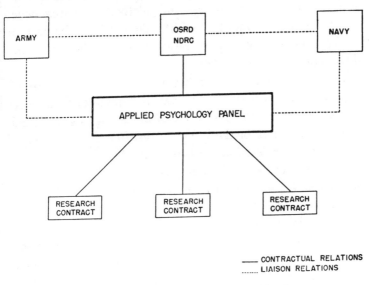

Figure 4. Organization chart of the Applied Psychology Panel.

ing of radar mechanics for the Signal Corps, was added to the group, at first as a Panel Member and shortly afterward as a full-time Technical Aide. The Panel held its first meeting, and the Committee on Service Personnel was dissolved on October 7, 1943.

The contract covering the Committee on Service Personnel was continued in force at the National Research Council and modified to support a Committee on Applied Psychology in the War. Stalnaker served as Chairman and Carmichael as Member. The new Committee supported a few preliminary studies requested by the Panel.

RESEARCH ADMINISTRATION

Despite the differences in organization, the scientific policies of the Committee on Service Personnel and the Applied Psychology Panel were the same, and the general relations to the initiation and conduct of research were closely similar. The typical pattern of activity and the policies of each may be described by reference to the work of the Committee alone. The pattern of Committee activity differed from the pattern of the Panel only in that no research was carried on by the direct employees of the Panel.

Most Committee research began through a request for assistance from the services. In a few instances it began on request from other divisions of the National Defense Research Committee. Later in the war new research often developed from previously established programs of the Committee. In every case the basic stimulus was a practical military need which could be satisfied more or less directly by psychological research. The Committee's emphasis was on the practical needs of the Army and Navy and on research which should not only meet these needs directly but should also furnish byproducts arising out of contact between Army and Navy officers and civilian psychologists. From the emphasis on practical contributions and service relations came a series of Committee policies which determined the character and location of the research projects.

It was a primary policy of the Committee to undertake no general research program unless asked to do so by the services in a formal written request through official channels. Only when a particular officer was able to put through such a request did the Committee feel assured that the program would have the support required to complete it and to secure general use of the results. While work often began in anticipation of a formal request, this occurred only when the Committee felt assured that the request was on its way through channels. For the same reason the Committee made no "grants-in-aid" to the several programs proposed to it by university psychologists. In a few instances, however, such programs were communicated to service representatives and aroused sufficient interest that formal service request for the work was made.

A second general policy was to insist that research projects be set up in field rather than in university laboratories. Thus all but a few of the research projects were conducted primarily in laboratories created by the Committee in Army and Navy establishments. It was the belief of the Committee that the supply of research personnel was so small, while practical needs were so great and so diverse, that the best use of every available research man was in the field in close contact with practical realities. Furthermore, the Committee believed that the problems of military psychology could best be solved by research on military personnel. Such research could be most useful if the investigator had an intimate knowledge of the conditions under which his results would be applied. Finally, it was the belief of the Committee that the services were making too little use of the elementary principles of psychology, as described in every freshman textbook. Greater practical values could be achieved by developmental research on the applications of elementary principles to specific problems than by the fundamental research typically carried out in university laboratories.

The policy of locating research laboratories at Army camps and Navy bases was reinforced by the by-products which the policy produced. The mere presence of a research staff interested in training, for example, stimulated interest in the training process among the local instructors and improved the quality of training. Psychologists who were concerned with classification suggested useful improvements in military equipment or operating procedures—if they became sufficiently familiar with the equipment and procedures in routine use. The research staffs educated local officers to think quantitatively about human problems; when this was done the problems often yielded to ready solution. And ideas for new research, better in terms of practical values than any which could be developed at a distance or on casual contacts, came from the research staffs and officers who watched the day-to-day progress of the work.

The policy was further reinforced by the fact that research in military establishments was readily accepted by the services as a valid reason for administrative change. The liaison officer felt that a proposed change was not the irresponsible suggestion of an outsider but the considered opinion of a man who was almost a part of the military organization. Suggestions supported by university-based research carried no such prestige.

Thus the original stimulus to research was always a practical need of the services and the usual source of inspiration to research was the services themselves. When the Committee agreed that the need had high priority and that psychological research might have a practical outcome, a survey of the scene was made.

Occasionally the survey indicated that research was not feasible. Frequently it developed that the services themselves, unknown to the original parties, had adequate agencies and personnel to handle the work. It was Committee policy always to encourage the development of service groups capable of handling research and to refuse requests if such development would be hindered by acceptance. Often it appeared that research by Committee psychologists could be carried on cooperatively with psychologists in the services and was actively desired by the service psychologists themselves. In these cases Committee or contractor's personnel were "attached" to the service laboratories. Occasionally the Committee had to establish a field laboratory of its own. Thus the survey generally ended in a research program, rather broadly defined, or in advisory activities which helped to meet the needs of the moment.

When the survey indicated the need for a major research project, the Committee formulated a broad program of work and a budget, and recommended a contractor for the consideration of the National Defense Research Committee. In all but two instances, the choice of a contractor was a simple reflection of the choice of a man to head the project. The employer of that man became the contractor and the man served as the "contractor's technical representative."

Choice of the contractor's technical representative was one of the Committee's most important functions. In the case of the first few projects, choice was simple and automatic. The Committee members themselves, as expected before the Committee was organized, took over the projects best suited to their backgrounds. In the later projects the man had to be located and persuaded to join in the work. It was necessary that the top man in each project be distinguished by ability and preparation for the particular job. It was necessary that he be a capable research administrator. It was necessary that he have a type of personality adapted to the minds and prejudices of service officers. It was usually necessary that he be relatively young in terms of the above characteristics, partly because of the physical strain involved in constant travel, partly because of the need for rapid decision and action, and partly because of the need for adaptability to new conditions. Since the Committee entered upon the scene at a relatively late date, most such psychologists were already in war service. A wise choice among the remainder was difficult, and as time passed it became more so. Near the end of the war new projects were staffed at the top chiefly by draining men from existing projects.

In the typical case the contractor developed one or more relatively independent laboratories and staffed these with research men drawn from the country at large. The laboratory staffs tended to develop their own programs. Committee supervision, once a project was definitely initiated, usually consisted chiefly in the positive suggestion of additional research fields, broadly defined, or the exercise of a mild veto power over new research and reports. Contractors' technical representatives were permitted the widest latitude in the initiation and conduct of new research and were entrusted with the major share of responsibility for liaison and practical application of results. As a result there were a few minor errors. These might have been avoided had the collective experience and judgment of the Committee members been made more effective. On the other hand a policy of close guidance and supervision would have led to very real difficulties. Close guidance by a committee always delays research. The result of the freedom granted to contractor's personnel was that current research topics occasionally surprised the Committee; even these cases were usually satisfactory to the services. The policy encouraged the development of a responsible professional group of military psychologists.

The freedom of the projects was partly a matter of deliberate policy and partly a matter of necessity. The necessity arose from the dispersed character of the research and the limited number of personnel available to the central administrative office of the Committee. In the typical case the original field laboratory of a project served as a research center and base of operations for work done at many outlying Army and Navy establishments. At the outlying establishments data were collected and referred back to the field laboratory for analysis and action. With the passage of time the work as a whole involved a large number of widely-separated locations over which close supervision would have been impossible without a very large supervisory staff.

The dispersed character of the work is clearly shown in the map of the United States which is reproduced in Figure 5. Here the locations of the contracting institutions and of the central laboratories at service installations are shown by capital letters. The many minor points at which work was also done are shown in capital and lower case letters. Only those points at which work was done by one or more staff members for a continuous thirty-day period are shown. Not shown are the large number of locations which were visited only for

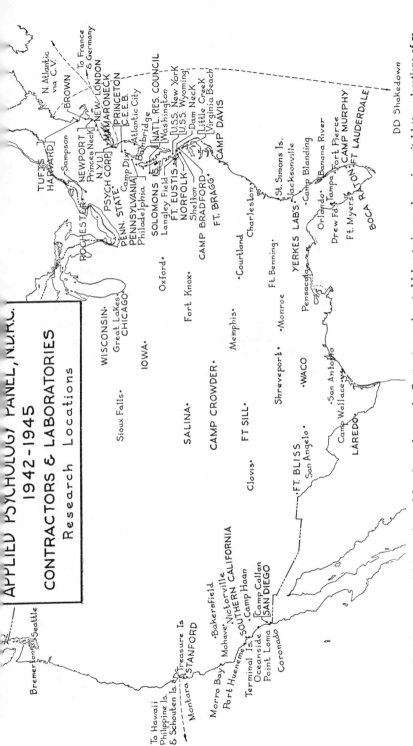

Figure 5. Contractors, central laboratories and locations of research. Contractors and central laboratories are shown in capital letters; locations of research appear in capital and lower case letters.

a few days in order to advise local authorities or to educate the research staffs themselves. Included under this definition, however, are the camps at which a quantity of data was collected. Also included are the Army or Navy centers to which representatives were attached, at least for 30 days, to the staff of some service commander in order to secure efficient use of research results.

The Committee was of assistance to its projects in a variety of ways. It located many of the staff members. It cut through the red tape which was inevitably involved in government-supported civilian research on military problems. The Committee secured clearance for staff members of the projects to receive information affecting military security and to enter service establishments. It gave such assistance as seemed appropriate in problems of draft deferment of research personnel. It secured priorities in obtaining equipment and in travel. It forwarded information relative to research under other organizations. These functions, in their totality, were absolutely required for successful operation of the projects. Only through its association with the National Defense Research Committee was the Committee able to perform them.

It has already been indicated that the contractor's technical representative was given the major responsibility for application of research results. The Committee shared in the process chiefly through its staff, who kept the liaison officers informed and at appropriate times introduced project research personnel to the liaison officers. Indirectly, however, the Committee staff was of major importance in the application of results through its work in the education of liaison officers in the value of psychological research. Without the active and informed assistance of the liaison officers in headquarters units, the work of the projects would have had but limited, local effectiveness.

LIAISON

From the start, as suggested in Chapter 1, the Committee enjoyed the closest liaison with the Navy. The Office of the

Coordinator of Research and Development of the Navy was charged with the function of general liaison with NDRC activities. This function was discharged with a broadly sympathetic and intelligent understanding which avoided innumerable pitfalls in the always difficult process of joint civilian-service operations. Psychological research concerned many sections of the Navy. It was new to the regular officers of these sections and it led to many proposals for the modification of traditional procedures. The civilian advisors and research workers were forced into an awkward position by the very nature of the situation. To the regular officer a researcher was like a fifth man at a bridge table, the "kibitzer" who sees much but has no responsibility for winning the hand. The wise kibitzer refrains from criticism and suggestions. It was the job of the psychologists to make suggestions. To make them positively rather than negatively was always a difficult task. To make them via a liaison officer relieved much of the personal strain, provided that the liaison officer himself understood the psychological as well as the Navy point of view. Commander McDowell and Captain Smith, respectively member and liaison officer of the Committee, steered the group through these troublesome waters.

Working together, Commander McDowell and Captain Smith drew the Committee into research on problems of importance to the Navy and saw to it that the research was performed at Naval stations where sympathetic officers were in charge. They saw to it that headquarters officers assigned for liaison duty on specific research projects were also understanding. This type of guidance allowed the research to start with a reasonable hope of success. The practical contributions of the research workers assured not only the continuance but the growth of liaison support.

In the Army a different type of liaison was enjoyed at the beginning of the work. Dr. W. V. Bingham was War Department representative and a member of the Committee. He was a psychologist and a civilian. His professional understanding and knowledge of psychological problems of interest to the Army were significant in educating the Committee and opening the doors to Army camps where research could most effectively be done. On the other hand, his duties for the Adjutant General were concerned only with classification, so that the Committee had no day-to-day contact with Army headquarters units responsible for training or new equipment.

With the passage of time the War Department Liaison Office with NDRC and the Committee staff grew into closer and closer contact. Capt. H. E. Clements, AUS, and Maj. H. E. Noble, AUS, came to give the Committee the same type of general assistance that was available from the start in the Navy. Dr. Don Lewis of the Office of the Chief Signal Officer also initiated several requests for research on training and new equipment and was a potent factor in the success of the resulting projects in the fields of communications and radar.

The liaison relations with the various branches of the Army and Navy grew more rapidly than the research. Psychological research proved to be of interest to an extraordinary variety of service organizations. As the early projects began to produce results the Committee was hard pressed to keep in touch with the many new liaison officers. The liaison officers were of two types—general and project. The general liaison officer was kept informed of the progress of all research so that he, in turn, could represent the interests of his branch of the service in the whole work of the Panel. The project liaison officer followed a single project only. The types of offices maintaining one or both types of liaison are illustrated in Table 1, which is a partial list only. In this table general liaison relations are shown in italics, and the name of the first general liaison officer to be appointed is given in each case. The list shows how widespread was service interest in the work of the Panel. Military psychology is significant to many types of service organizations.

TABLE I. Partial list of service offices maintaining formal liaison with the Applied Psychology Panel. General liaison relations are shown in italics to distinguish them from liaison with one or a few projects. The list gives the name of the first general liaison officer to be appointed from any office.

NAVY LIAISON

NAVY DEPARTMENT OFFICES
Office of the Coordinator of Research and Development (Capt. L. P. Smith, USN)
Office of the Commander in Chief, U.S. Fleet
 Readiness Division (Comdr. P. E. McDowell, USN)
Bureau of Aeronautics
Bureau of Medicine and Surgery (Lt. Comdr. J. G. Jenkins, USNR)
Bureau of Naval Personnel
  Administrative Division
  Enlisted Personnel Division (Lt. Comdr. H. J. Pohl, USN)
  Planning and Control Division (Lt. Comdr. J. G. O'Brien, USNR)
  Quality Control Section (Lt. Comdr. C. M. Louttit, USNR)
  Research (Lt. Comdr. R. N. Faulkner, USNR)
  Standards and Curriculum Section (Lt. Comdr. A. C. Eurich, USNR)
  Training Division (Capt. F. U. Lake, USN)
Bureau of Ordnance
Bureau of Ships
Bureau of Supplies and Accounts
Deputy Chief of Naval Operations (Air)
Interior Control Board

FIELD OFFICES
Advanced Fire Control Schools, Washington, D.C.
Amphibious Training Command, Atlantic Fleet
Amphibious Training Command, Pacific Fleet
Chelsea Naval Hospital, Boston, Mass.
Fleet Training Command, Seventh Fleet
Free Gunnery Standardization Committee
Naval Air Station, Banana River, Fla.
Naval Air Technical Training Command
Naval Radar Training Station, St. Simon, Ga.
Naval Training Station, Newport, R.I.
Operational Training Command, Atlantic Fleet
Operational Training Command, Pacific Fleet
Staff, Atlantic Fleet
U.S. Submarine Base, New London, Conn.
 Medical Research Laboratory (Comdr. C. W. Shilling, USN, MC)

MARINE CORPS
 Division of Aviation (Maj. G. M. Morrow, USMC)
 Personnel Department (Lt. Col. H. E. Dunkelberger, USMC)

ARMY LIAISON

WAR DEPARTMENT OFFICES

War Department Liaison Office with NDRC (Capt. H. E. Clements, AUS)
War Department General Staff, G-1 (Col. Edmund Lynch, USA)
Adjutant General's Office (Dr. W. V. Bingham)
Assistant Chiefs of Air Staff
    Materiel and Supplies
    Operations, Commitments, and Requirements
    Training
Headquarters, Army Ground Forces (Capt. D. L. Emmel, AUS)
Office of the Air Surgeon (Lt. Col. J. C. Flanagan, AUS)
Office of the Chief of Ordnance
Office of the Chief Signal Officer (Dr. Don Lewis)
War Department Special Staff
    Information and Education Division

FIELD OFFICES

Air Technical Service Command, Wright Field
    Aero-Medical Laboratory
    Armament Laboratory
Antiaircraft Artillery Board
Antiaircraft Artillery Command
Antiaircraft Artillery School, Camp Davis
Antiaircraft Artillery School, Fort Bliss
Armored Medical Research Laboratory, Fort Knox
Army Air Forces Proving Ground Command
Army Air Forces Training Command
Central School for Flexible Gunnery, Laredo Army Air Field
Central Signal Corps School, Camp Crowder
Field Artillery Board
Field Artillery School, Fort Bragg
Field Artillery School, Fort Sill
Second Air Force
Signal Corps Engineering Laboratory, Bradley Beach, N.J.
Signal Corps Ground Signal Agency
Signal Laboratory, Camp Evans, N.J.
Signal Security Agency

MISCELLANEOUS

British Admiralty Delegation
War Shipping Administration

PROJECTS AND CONTRACTORS

The formal projects began with the two problems which
had been of concern to the Navy liaison officers at the time of

formation of the Committee on Service Personnel. The two projects were named Research and Development of the Navy's Aptitude Testing Program, and Selection and Training of Naval Gun Crews. The first was assigned to J. M. Stalnaker under a contract with the College Entrance Examination Board. M. S. Viteles was placed in charge of the project on Naval Gun Crews under a contract with the University of Pennsylvania. In each case work began on September 1, 1942.

As the first two projects were being established, consideration was given to a third. Division 7 of the National Defense Research Committee asked the Committee on Service Personnel to continue research begun by Division 7 on the selection and training of heightfinder operators. The request was accepted, and on October 1, 1942, C. H. Graham began the work under a contract with Brown University.

These three original projects provided the prototypes for nearly all the succeeding projects. Following the research on the Navy aptitude tests came a series of projects and subprojects on other tests and on classification procedures. In this series the research interest was limited to the improvement of classification, but the tests and procedures under consideration were intended for all personnel. Following the research on Navy gun crews there came a series of projects on the selection and training of various kinds of specialists. In most instances there was a greater emphasis on training than on selection, but in all cases a single type of specialist received the concentrated attention of an entire project. In the heightfinder project the study of aptitude and training was closely tied to research on the design and operation of military equipment. The work of this project led directly to a series of projects on the control of gunfire in which the equipment element played an important part.

In Table 2 the list of major projects is shown, together with the psychologist originally placed in charge of the research, the name of the contractor, and the initial and terminal dates of the contract. First in the list is placed the Committee on

TABLE 2. The list of the major research projects of the Applied Psychology Panel showing the topic, the psychologist in charge, the contractor, and the initial and terminal dates of the contract.

Topic	Psychologist in Charge	Contractor	Dates	
			INITIAL	TERMINAL
I. GENERAL AND ADMINISTRATIVE				
The Committee on Service Personnel. The Committee on Applied Psychology and the War	L. Carmichael	National Academy of Sciences	June 20, 1942	Oct. 31, 1945
II. CLASSIFICATION				
Navy Aptitude Tests	J. M. Stalnaker	College Entrance Exam. Board	Sept. 1, 1942	Oct. 31, 1945
Emotional Stability	C. H. Graham	Brown University	Oct. 1, 1942	Apr. 30, 1945
U.S.S. New Jersey	J. M. Stalnaker	National Academy of Sciences	—	—
Interest Test	T. L. Kelley	Harvard University	Dec. 1, 1943	Dec. 31, 1944
Leadership	H. E. Garrett	National Academy of Sciences	Dec. 20, 1943	June 8, 1944
Aids to Classification	R. K. Campbell	Stanford University	Apr. 1, 1944	Aug. 31, 1945
III. SELECTION AND TRAINING OF SPECIALISTS				
Naval Gun Crews	M. S. Viteles	University of Pennsylvania	Sept. 1, 1942	Oct. 31, 1945
Radio Code Operators	G. K. Bennett	Psychological Corporation	Oct. 1, 1942	Oct. 31, 1945
Radar Operators	D. B. Lindsley	Yerkes Laboratories	Mar. 15, 1943	Sept. 30, 1945
Voice Communication	G. K. Bennett	Psychological Corporation	Apr. 7, 1943	Aug. 31, 1945
Night Lookouts	C. H. Wedell	Princeton University	Sept. 1, 1943	Dec. 1, 1944
Engineering Crews	M. S. Viteles	University of Pennsylvania	Oct. 1, 1943	Oct. 31, 1945
Amphibious Personnel	K. R. Smith	Pennsylvania State College	Feb. 24, 1944	June 30, 1945
Cargo Handlers	F. L. Ruch	University of Southern California	July 1, 1944	Feb. 28, 1945
IV. CLASSIFICATION, TRAINING, AND EQUIPMENT IN THE CONTROL OF GUNFIRE				
Heightfinder Operators	C. H. Graham	Brown University	Oct. 1, 1942	Dec. 31, 1943
Rangefinder and Radar Operators	C. H. Graham	Brown University	Sept. 30, 1943	Dec. 31, 1943
	W. J. Brogden	University of Wisconsin	Jan. 1, 1944	Jan. 31, 1945
Antiaircraft Battery	L. C. Mead	Tufts College	June 1, 1943	Sept. 30, 1945
Lead Computing Gunsights and Directors	C. H. Graham	Brown University	Dec. 1, 1943	Sept. 30, 1945
Field Artillery Gunsights	L. C. Mead	Tufts College	May 25, 1944	Aug. 31, 1945
B-29 Gunnery	W. J. Brogden	University of Wisconsin	Sept. 1, 1944	Oct. 31, 1945
B-29 Gunnery Training	R. R. Sears	University of Iowa,		

Service Personnel, since this itself was a project. Next come the projects dealing with classification alone; within this series the listing is chronological. Following, again in chronological order, are the projects on the selection and training of specialists. At the bottom appears the list of projects which studied problems of classification, training, and equipment in the control of gunfire.

In the remaining chapters of this book, which describe the Panel's research in relation to military problems, the chronological sequence of development will usually be ignored. The chronological sequence was not logical because the war situation prevented an orderly development. Practical and immediate demands were too great to permit a planned rational sequence. Research was broken off again and again, for instance by more urgent needs for the soldiers or sailors who were serving as the "guinea pigs" in Panel experiments. In retrospect it is clear that considerable time would actually have been saved the Army and Navy if men had been assigned to duty as subjects of research and if no deviation from the assignment had been allowed, but most of the work took place before the value of psychological research in the military situation had been overwhelmingly proved. It was not until near the end of the war that proof in quantity was available. At this time too, the manpower strain was eased and psychological experimentation became relatively easier. Then the orderly processes of science became more common.

As a result of the difficulties of research in the early days of the war, recommendations for action were often made before the complete evidence was obtained. Had it been otherwise the work would have had little practical value in the emergency. The good judgment and technical competence of the Panel representatives was proved later, sometimes long after the recommendations were put into effect. That the scientific judgment was good was often proved by others as well as by the Panel itself. That the judgment of the group was sound in terms of the wartime realities was proved by the steady increase in service requests for research. The following chapters sample some of the work and give the evidence of its value.

CHAPTER 3

RESEARCH ON CLASSIFICATION

AT the beginning of a war when mobilization occurs, manpower problems reach a peak. As industry turns to war production and demands that every capable man be kept on his job, the services expand. New recruits are inducted. The men of the regular Army and Navy are thinly spread through training camps and combat zones and the ranks are filled with civilians in uniform, men who know little or nothing of the art of war.

Consider the concrete problem facing the captain of a new battleship, who, in wartime, must take his ship away from a pier and after a short shakedown cruise in quiet waters sail away to combat.[1] Early in 1943 the U.S.S. *New Jersey* was approaching completion in a Navy Yard. She consisted of a steel shell enclosing intricate mechanical and electrical devices. The devices were the finest that modern science could produce. It was possible for these devices to detect enemy ships and planes miles away. It was possible for these devices to place tons of high explosive within a few yards of any point over a wide radius. It was possible for this pride of the Fleet to serve as a nerve center for a whole task force. Yet who would see to it that the enemy was detected? Who would aim and fire the guns? Who would maintain effective communication with the other ships of our Fleet? The answer had to be the 2,600 sailors of whom roughly half were raw recruits and more than two thirds had never before been to sea.

As the date for the commissioning of the *New Jersey* approached, her crew began to arrive. At first they came from recruit camps, training centers, and other ships in little groups of two or three. Later, they arrived in larger and larger groups until on one week-end over 1,000 appeared. Spread among the

[1] The following paragraphs are based on STR I, Chapter 11 by John L. Kennedy and Chapter 12 by Norman Frederiksen.

inexperienced was a thin nucleus of "veterans," men with the experience of at least one sea cruise, and a very few regulars, men trained to the sea. Of the veterans and regulars some had seen battle. It was the problem of the captain, and the direct responsibility of his executive officer, to weld the men and the steel ship into a fighting unit. A few weeks were available at the pier and a few months in a quiet shakedown cruise. Every possible moment had to be used to best advantage.

In 1943, in the routine course of events, the captain and his executive officer had help from the industrial technicians who had produced the *New Jersey's* equipment. They had help from the Navy itself which sent its specialists on guns, radar, communications, engines, and other equipment. In the routine course of events they would have had little technical help on their manpower problem, but in the case of the *New Jersey* the executive officer, Commander McDowell, had been closely associated with the development of a psychological section in the Bureau of Naval Personnel and with the development of the Committee on Service Personnel. He appealed to these organizations for help and offered them an opportunity to observe the application of their techniques in a relatively small-scale situation.

The response of the Bureau and Committee was to organize a classification system for the *New Jersey*.[2] As the crew members arrived they were tested for verbal, mechanical, arithmetical, and clerical aptitude; for emotional stability; for ability to speak clearly over a telephone and to understand telephonic communications in noise; for visual acuity, stereoscopic acuity and night visual ability. As rapidly as the test scores became available the men were interviewed; their civilian experience, hobbies, and interests were determined and their previous Navy experience was evaluated. The inter-

[2] For the Bureau of Naval Personnel, Lt. Comdr. A. C. Eurich, USNR, was chiefly responsible; for the Committee, J. M. Stalnaker was in charge. Ensign (later Lt.) H. A. Black, USNR, as aide to the executive officer, carried through the program with the help of many representatives of the Bureau, Committee, and other cooperating agencies.

viewer considered the fitness of each man for the many specific duties on board ship and assigned each man accordingly.

The results of the whole program were striking. They cannot be evaluated in the scientific sense,[3] but the experience gained by all concerned was effective in guiding later work. In addition the whole classification program of the Bureau of Naval Personnel gained support in Navy quarters where support was most effective. The captains of ships adjacent to the *New Jersey* were at first quite skeptical. As the program materialized they began to demand the same service for their ships. A few months later Capt. C. F. Holden, USN, the skipper of the *New Jersey* herself, said:[4]

"The initial placement of personnel has definitely contributed to the apparent extraordinary rapid progress made during the shakedown period by the crew of this vessel. By this method, the following benefits were derived:

a. Obviously unsuited personnel were eliminated from the start without waste of training effort.

b. Transfers between divisions and departments were kept at a minimum due to proper placement prior to the training period.

c. There was a reduction in the number of men failing to profit by training due to a reduction in the malplacements."

The problems of the U.S.S. *New Jersey* reflected, in minia-

---

[3] H. A. Black attempted to follow up by a validational study of the tests and procedures. As a criterion he used rating scales of performance of the men on board ship. The samples for specific duties were small and the rating scale was general. Low positive correlations were obtained between test scores and performance ratings. These had little statistical significance. Black's report suggests the need for tests directed toward combat as well as toward school aptitudes. It emphasizes the value of precise information on civilian experience and interests. Lt. H. A. Black, USNR, Memorandum to the Commanding Officer, U.S.S. *New Jersey*, Subject: *Personnel Classification on Board Ship—Report of*. April 14, 1945. Forwarded under covering Memorandum of May 18, 1945, BB62/P17–2, Serial 224 from the Commanding Officer, U.S.S. *New Jersey* (BB62) to the Chief of Naval Personnel.

[4] Capt. C. F. Holden, USN, Memorandum from Commanding Officer to Chief of Naval Personnel, Subject: *Organization of Ship's Company*. Reference BB62/P16–3, Serial 570. August 31, 1943.

ture, the problems of mobilization everywhere. At Army and Navy centers over the country many hundreds and even thousands of recruits arrived weekly. Some were Ph.D.'s, while others had a few years of grammar school; some had genius, while others were feebleminded; some were highly skilled mechanics, engineers and instrument makers, while others had no special skill. Some said that they had one or more of these characteristics, while others were inarticulate about themselves. It was the problem of the classification officer to sort out the men. Within a few days the men had to be assigned to duty or to training for special duty. Assignment might be based on time of arrival, on the alphabet, or on chance; or assignment could be based on knowledge of a man's aptitudes and other characteristics. From the beginning of the war to the end, a major effort of the research psychologist was to develop tests and procedures which would facilitate the systematic assignment and reassignment of men.

In the Navy the following system of classification was developed:

1. Each recruit was tested and information about his characteristics was collected from various sources. The following information became available:
   a. Aptitude test scores.
   b. Data from the medical examination.
   c. Educational and occupational data derived from a form which was completed by each recruit himself and which covered many items of biographical information.
   d. Navy record, if any.
2. Each recruit was interviewed. The interviewer was furnished the information listed above and had at hand:
   a. A list of Navy jobs for which personnel was required at the moment.
   b. Tables showing the relation between Navy jobs and civilian jobs.
3. The interviewer, by talking with the sailor, checked the

accuracy of the information on the educational and occupational record, extended it as needed, and tried to obtain a more precise idea of the nature of any previous Navy experience.

4. The interviewer tried to learn the nature of the sailor's interests. The recruit had previously seen a motion picture describing various types of Navy duties; he indicated those duties which he thought might interest him. The nature of hobbies and the degree of special attainments were considered by the interviewer.

5. The sailor's characteristics were related to the list of Navy duties. Frequently the interviewer was relatively inexperienced with the Navy himself, but he had studied verbal descriptions of Navy duties and had before him the tables which showed the relation of civilian jobs to Navy jobs.

6. The interviewer made a first recommendation and a second recommendation of the specific duty for which the sailor was best fitted.

7. The interviewer judged the quality of the sailor for special school training as (a) exceptionally well qualified for the specific recommended duty, (b) well qualified in general but lacking a "definite pattern of qualification," (c) lacking high qualifications but reasonably qualified if the manpower need were great, and (d) lacking in qualifications for special school training.

8. The interviewer's judgments and any new information obtained were entered in the records.

The sailor was assigned to actual duty on shipboard by ships' officers who studied the records, sometimes with the help of a classification officer. The sailor was assigned to school training by mechanical card-sorting systems. In either case the interviewer's recommendations and judgments of quality were the basic determining factors if manpower needs were not overwhelming and if time permitted. The interviewer's functions were complex. They had to be exercised repeatedly, hour after hour, day after day, as successive men were interviewed every ten to twenty minutes. It was the task of the research psychologist to improve the interviewer's work by furnishing him with more, and more accurate, information about each recruit, providing a wider range of relevant information, and by reducing the need for vague judgments based only on non-quantitative information.

Three approaches to the improvement of the classification program were undertaken by the Applied Psychology Panel: (1) The development of tests of relatively generalized human capacities and traits. Verbal, mechanical and numerical aptitudes are required in varying degree for good performance in many parts of the services. So, too, are the capacity to lead men in combat, the degree of interest in various kinds of activities, and emotional stability. In each case the Panel undertook direct study of the problem. (2) Studies of the administration of a classification program. Panel research included evaluations of the success of classification and the development of aids and training methods for classification interviewers. (3) The development of several tests of the narrow specialized aptitudes and skills particularly required in one or a few jobs. Panel research on tests for specific aptitudes was included in a series of more general psychological studies of special groups of personnel. The development of tests of the general aptitudes and research on the classification process will be described in this chapter. The research on special tests will be described in later chapters.

The Navy Basic Classification Test Battery[5]

The most significant Panel research on general tests and the process of classification was carried out in a project entitled Research and Development of the Navy's Aptitude Testing Program.[6] The project was originally requested by

[5] This section is based on *STR* I, Chapter 2 by Dael Wolfle and Chapter 14 by Norman Frederiksen.

[6] A summary and complete bibliography of the work of the project are con-

the Bureau of Naval Personnel to study aptitude tests and classification procedures for enlisted men and officers. Later an achievement and proficiency testing program was developed, which will be described in Chapter 8. J. M. Stalnaker had general charge of the work, and Harold Gulliksen developed the program in detail. The research staff is listed in Table 3.

TABLE 3. The research staff of the Panel project on Navy aptitude tests.

Contractor: The College Entrance Examination Board
Contractor's Technical Representatives: J. M. Stalnaker and Harold Gulliksen
Project Director: Harold Gulliksen
Staff: M. D. Bown, H. S. Conrad, T. L. Engle, N. A. Fattu, Norman Frederiksen, C. M. Harsh, L. D. Mays, D. A. Peterson, G. A. Satter, M. E. Thompson

From the beginning of project work on September 1, 1942, to the end of World War II the project worked in close cooperation with the Bureau of Naval Personnel.[7] In most cases the relationship was so close that it is difficult to assign responsibility for results to one group or the other. Mutual cooperation in planning, in the assignment of part tasks, in the conduct of the work, and in the evaluation of results was the rule. The members of the project staff essentially served as a part of the Standards and Curriculum Division of the Bureau of Naval Personnel. Their freedom to do so illustrates the flexi-

tained in Herbert S. Conrad, *Summary Report on Research and Development of the Navy's Aptitude Testing Program: Final Report on Contract OEMsr-705 (September 1, 1942—October 31, 1945).* OSRD Report 6110. October 31, 1945. College Entrance Examination Board. Washington, D.C., Applied Psychology Panel, NDRC. A more complete account of research by and for the Bureau of Naval Personnel appears in Dewey B. Stuit, Editor, *Personnel Research and Test Development in the Bureau of Naval Personnel.* Princeton, N.J., Princeton University Press. 1947.

[7] Lt. Comdr. A. C. Eurich, USNR, Lt. R. N. Faulkner, USNR, and Lt. G. L. Bond, USNR, were the officers most generally concerned with the work of the project.

bility of the organizations under the National Defense Research Committee.

To replace the aptitude tests in use in the Navy in the fall of 1942, the Bureau of Naval Personnel and the Panel developed a series of new tests for enlisted personnel. By the winter of 1943 the new tests had been prepared, standardized, and put to use. Collectively they were named the U.S. Navy Basic Classification Test Battery. Validation of the Basic Battery followed its development and continued throughout the war as new forms of the tests were developed.

It was originally decided that specialist training in the Navy required the recruit to possess verbal, numerical, or mechanical aptitude. In some Navy schools only one of the three aptitudes might be required; in most, two or even all three might be desirable. Thus a test battery was required which would include at least one test for each aptitude. The project developed the verbal and numerical portions;[8] the Bureau developed the mechanical portions.

The tests of the Basic Battery are described in Table 4, which names the tests and component subtests, shows the number of items and time limits for each, and gives typical items from each test or subtest. The General Classification Test (GCT) and the Reading Test were designed to measure verbal aptitude. The purpose of the other tests is defined by their names—Arithmetical Reasoning, Mechanical Aptitude, and Mechanical Knowledge. The last gave two scores, one stressing electrical information and one stressing mechanical information.

In order to give the reader some insight into the type of men who made up the enlisted personnel of the Navy the difficulty of each item listed in Table 4 is shown at the right of the item. The difficulty of an item is given by the ratio of the number

[8] Norman Frederiksen, *Preparation of the United States Navy General Classification Test—Form 1 and the United States Navy Tests of Reading and Arithmetical Reasoning—Form 1*. Project N-106, Memorandum No. 7. June 23, 1943. College Entrance Examination Board. Washington, D.C., Applied Psychology Panel, NDRC.

TABLE 4. The nature of the tests of the Basic Battery. For each test or subtest, two examples are given. The number to the right of each example gives the proportion of a national sample of 500 recruits who passed the item if they attempted it at all. The tests marked with an asterisk were developed under the Panel.

*I. General Classification Test (abb.: GCT)

    a. Sentence Completion (30 items, 10 minutes). Which one of the five words best fits in the sentence:

        Always . . . . . the salute of those under you.   .89
            1. approve 2. seek 3. appreciate 4. watch
            5. return

        It was clear in 1942 that victory over Japan would be an . . . . . victory indeed if it were coupled with a United Nations defeat in Europe at the hands of Germany.   .16
            1. important 2. appalling 3. empty 4. officious
            5. indirect

    b. Opposites (30 items, 10 minutes). Find the one word which means the *opposite* of the word in capital letters:

        ADEQUATE   .56
            1. improper 2. deceitful 3. insufficient
            4. unprotected 5. unnecessary

        ACCELERATE   .48
            1. punish 2. grovel 3. release 4. soothe
            5. retard

    c. Analogies (40 items, 15 minutes). Select the one of the five words which best completes the thought:

        Ship is to *lifeboat* as plane is to   .68
            1. safety belt 2. hangar 3. aircraft carrier
            4. glider 5. parachute

        Crest is to *trough* as hill is to   .42
            1. mountain 2. meadow 3. slope 4. lake
            5. valley

*II. Reading Test (abb.: READ) (30 items, 25 minutes). Read the following paragraph and then complete the sentences below it:

Cruisers have light armor, carry guns of moderate size, and are able to travel at high speed. Cruisers whose largest guns are greater than 6 inches are known as heavy cruisers, while those whose largest guns are 6 inches or less are known as light cruisers. All cruisers have very large fuel tanks in order to maintain high speed for a long time. A cruiser is divided into numerous watertight compartments, so that a hole in one part of the ship will flood only a part of the ship.

Cruisers whose largest guns are 6-inch guns are .72

a. light cruisers
b. medium cruisers
c. heavy cruisers
d. auxiliaries
e. not described in the paragraph

Cruisers are able to maintain high speed for a long time because .89

a. they have a wide beam
b. they are divided into many small air-filled compartments
c. they have only three gun turrets
d. they carry a large supply of fuel oil
e. they do not have armor plate

*III. Arithmetic Reasoning Test (abb.: ARI) (30 items, 30 minutes)
Solve each problem and indicate the correct answer:

In an armor piercing shell 3 per cent of the total weight of the shell is bursting charge. What is the weight, in pounds, of the bursting charge in a 2100 pound shell? .49

a. 7 b. 63 c. 300 d. 630 e. 700

If 3 parts sand by weight are mixed with 4 parts cement, what fraction of the total dry mixture is sand? .42

a. 3/7 b. 4/7 c. 1/4 d. 3/4 e. 4/3

IV. Mechanical Aptitude Test (abb.: MAT)
a. Block Counting (45 items, 6 minutes). The sample item shown and answered for all candidates was: Count the number of blocks that touch a block with a given letter on it and check the number opposite the letter below.

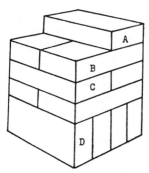

A: 2 3 4 5 6
B: 2 3 4 5 6
C: 2 3 4 5 6
D: 2 3 4 5 6

b. Mechanical Comprehension (44 items, 20 minutes). Which one of the three answers to the question about the picture is correct?

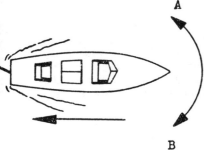

Which direction will the bow swing if this ship is going in reverse?

(1) Toward A.   .51
(2) Toward B.
(3) The bow will not swing either way.

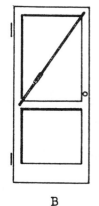

Which is the better way to brace a screen door to keep it from sagging?

(1) A.   .57
(2) B.
(3) The bracing is equally good.

A   B

c. Surface Development (40 items, 8 minutes). Below is a PATTERN and a PICTURE of the figure which can be made from this PATTERN. Side x in this PATTERN corresponds to side x in the PICTURE. Notice that each numbered line in the PATTERN is the same as one of the lettered edges in the PICTURE. You are to find the letter in the PICTURE which corresponds to the number in the PATTERN.

SIDE (x) IN THE PICTURE IS MARKED IN THE PATTERN

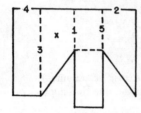
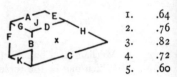

I.	.64
2.	.76
3.	.82
4.	.72
5.	.60

PATTERN   PICTURE

V. Mechanical Knowledge Test. In this test there are 135 items and the total time allowed is 37 minutes. Of the items, 75 relate to mechanical knowledge and 60 to electrical knowledge. In each case there are pictorial items and verbal items. Both an electric score (abb.: MK[E]) and a mechanical score (abb.: MK[M]) are obtained.

a. Mechanical (Pictorial). What lettered picture goes with the picture at the left?   .60

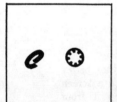

b. Mechanical (Verbal). Complete the sentence.
In a gasoline engine the gas mixture should explode in   .58
    a. the cylinder
    b. the intake manifold
    c. the exhaust manifold
    d. the carburetor

c. Electrical (Pictorial). What lettered picture goes with the picture at the left? .48

 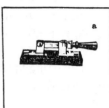  

d. Electrical (Verbal). Complete the sentence.
The metal used in the core of an electromagnetic coil is .27
    a. iron b. copper c. lead d. brass

of men answering the item correctly to the number attempting it; thus, the lower the figure shown, the greater is the difficulty. The difficulties listed were obtained on a national sample of recruits in 1944. The items shown varied in difficulty from very easy to very hard. It is interesting to note, in these technical days, that over half the wartime Navy recruits could not give the opposite of the word accelerate, calculate a simple percentage, or formulate a fraction. On the other hand nearly ninety per cent knew that the thing to do to a salute was to return it. Of those who failed the last-mentioned item in 1944, about half said that the appropriate response is to appreciate the salute. Although one may wonder about the attitude of those who checked this particular answer, the question was a good one. Men who failed it did poorly on the subtest as a whole.

The most important question one can ask about an aptitude test is whether it is valid. This is to ask whether it does in fact predict performance in school or on the job. A test may measure a human characteristic with great accuracy and precision. Accuracy and precision are irrelevant unless the characteristic is important on the job. The importance of the characteristic for a particular job can only be demonstrated by proving that the test predicts how well the job will be done by particular men who have particular scores on the test. The question is a relative one of course. To be useful the test need only add a significant amount to the accuracy of prediction provided by the existing method of assignment, whatever the existing method may be. The test should decrease the error of assignment of a chance system, or of a system based on the alphabet, on records of experience, or on pre-existing aptitude tests.

The tests of the Basic Battery were designed to choose the men of higher qualifications for Navy Schools. Their value for this purpose can be estimated by determining the degree to which men who obtained a high, medium, or low score on any particular test later received high, medium, or low marks, respectively, in one or more schools.

The accuracy of prediction is commonly measured in terms of the coefficient of correlation (r) between the test scores and the school marks, or criterion scores. The higher the correlation, the better is the test. In general, validity correlations of .50 and above are very satisfactory since chance factors, variations in training methods, or changes of motivation always reduce validities well below the perfect prediction represented by a correlation of 1.00. In dealing with large masses of men, as in the military situation, tests with validities of the order of .20 or lower may be useful. One's attitude toward a test of any given validity must be qualified by a number of considerations some of which will be discussed below, for example, by the character of the criterion, and the independence of the test from other tests included in the same battery.

In one early study of validity,[9] the Bureau of Naval Personnel and the Panel project computed the correlations between each of the six scores given by Form 1 of the Basic Battery and school grades in standard Navy training for each of thirteen specialties. The results are shown in detail in Table 5, which also gives the best test combinations and the validities of the combinations. A total of seventy-four correlations were computed for single tests. Of these, twenty-two were

---

[9] Enlisted Classification Bulletin, September 1945, p. 8. Washington, D.C., Bureau of Naval Personnel.

between .50 and .70, and twenty more were between .40 and .50. The best test for each school gave a correlation greater than .50 in eight of the thirteen schools and in no case was the best test below .39. As shown in the last column of the table, combinations of the best two tests gave predictions above .50 in all but two of the schools. For Form 2 of the Basic Battery, a similar study showed that the best single test for each school

LE 5. Mean correlations between scores on the Basic Battery tests, Form 1, and final les in thirteen elementary service schools (correlations corrected for curtailment in ge of talent). The abbreviations of the names of the tests are explained in Table 4. stands for Women's Reserve.

chool	N	GCT	READ	ARI	MAT	MK(M)	MK(E)	Best test combination	Multiple r
tion ord-nce men	184	.63	.59	.59	.54	.35	.42	GCT & ARI	.66
c engi-ering	1,480	.52	.52	.63	.52	.46	.39	ARI & MK(M)	.66
el	2,160	.42	.35	.36	.26	.43	.46	GCT & MK(E)	.53
trical	1,747	.52	.52	.59	.44	.35	.49	ARI & MK(E)	.63
control-en	198	.52	.56	.49	.36	.18	.07	GCT & READ	.58
ner's ates	1,677	.38	.39	.31	.28	.40	.43	READ & MK(E)	.49
pital rps	449	.50	.45	.36	.27	.22	.30	GCT & READ	.51
hinist's ates	755	.33	.27	.44	.48	.48	.45	MAT & MK(E)	.54
ar perators	1,053	.60	.67	.61	.50	.35	.38	READ & ARI	.70
al	984	.49	.43	.44	.25	.13	.13	GCT & ARI	.52
ekeep-s (WR)	678	.44	.47	.59	.37	...	...	READ & ARI	.60
pedo-en	880	.32	.35	.28	.27	.39	.35	READ & ARI	.44
men WR)	738	.62	.59	.63	.38	...	...	GCT & ARI	.67

had a validity coefficient above .50 in ten out of the twelve cases studied.

These correlations were high. They may be contrasted with comparable correlations for the old Navy tests which were in use before the new Battery was constructed. Direct comparison on the same population was not possible, but a study similar to the above was carried out on the value of the old tests for six comparable Navy schools. Of a total of thirty correlations, four were above .50 and three were between .40 and .50.[10] The new battery was a marked improvement over the old.

In the development of a modern aptitude test battery a series of technical procedures are followed. First it is necessary to determine the purpose of the tests, to determine just which critical skills and abilities are to be measured. Second, it is necessary to have a general notion of the characteristics of the men to be tested and to be acquainted with the conditions and limitations of the situation in which the men will be tested and the results used. When the situation has been clearly defined, it is customary to prepare an experimental form of a test, to administer it to a reasonable sample of the group, and to perfect the experimental tests as measuring instruments. When all these preliminaries have been cleared away, a trial form of the battery can be issued with a reasonable expectation that it will be useful. The usefulness is determined by studies of validity. Finally, a continuing program of research is required to insure that the battery continues to be useful as conditions change.

In a normal peacetime development the steps outlined above would be taken in order. Each step would require its own definite program and, for each, special techniques would be used. For greatest efficiency a large-scale permanent professional organization is required. In the hasty creation of a

[10] $N=105$ to 234. The course grades had a reliability above .90 in all six schools. Norman Frederiksen, *Validity of Navy Aptitude Tests in Service Schools at the Great Lakes Naval Training Station.* OSRD Report 3245. January 31, 1944. College Entrance Examination Board. Washington, D.C., Applied Psychology Panel, NDRC.

test program for wartime use, it was necessary to create the research staff, to carry out as much research as possible, but above all to put the best available procedures in use at once. The degree of judgment and the technical skill of the men of the Bureau of Naval Personnel and the Panel can be evaluated in the validity coefficients given above for the first form of the Basic Battery. Improved results appeared later as the more orderly processes of science replaced the haste of the first months.

The preliminary definition of purposes required a decision as to whether the test battery should be aimed at predicting success in service schools or eventual success in military jobs in combat and elsewhere. For two reasons the prediction of school success was chosen as the purpose of the tests of the Basic Battery. First, it was important in itself to prevent the waste of time involved in training men who were poorly qualified and whose very presence in training school lowered the level of the training which could be given. Second, it was possible to obtain measures of success in school in the form of grades, promotions to petty officer status, and failure or reassignment at the end of training. Such measures were required to validate the tests. Measures of success in military jobs, and particularly measures of combat success, did not exist at the time the work began. And in so far as the service schools really fulfilled their function of preparation for military duty, the prediction of school success constituted effective prediction of success on the job.

Having determined that the function of the tests was to predict school performance, information was sought on the psychological characteristics required for success in the many types of service school. These characteristics were given preliminary definition by a careful job analysis. The nature of a formal job analysis will be described in Chapter 4 in relation to the Panel's studies of training. For classification purposes, informal job analyses defined the types of items to be used in the test batteries. The same type of information was sought in detailed statistical analyses of the characteristics of the aptitude tests previously used in the Navy.[11]

Although the prediction of school success was taken as the function of the tests, it was necessary to evaluate the various training programs themselves. Training which was unrealistic would have provided misleading information on the value of tests. Some schools were too "academic," i.e. too verbal or theoretical; others used examinations which were unreliable or unrelated to actual training; still others were allotted insufficient time and facilities to bring trainees up to a stable level of performance. In such cases the criterion of school success was not a meaningful measure against which to check test effectiveness. As a result of increasing attention to the training program in relation to test evaluation, a program was established to improve service school examinations. To this program many contributions were made by the Panel project on Navy aptitude tests (see Chapter 8).

The validity data shown above in Table 5 illustrate several concrete cases in which service school practices probably produced misleading evidence on test effectiveness. For the Basic Engineering and the Electrical schools, the Arithmetical Reasoning Test should not have been the best predictor of school grade; these schools were training for duty as a mechanic or as an electrician, and arithmetical ability should have been less important than mechanical ability. In the Gunner's Mate and the Torpedoman schools, likewise, the verbal tests should have been poorer predictors than the mechanical tests. It seemed likely that the curricula and the examination practices of those schools, rather than the characteristics of the aptitude tests, produced the unexpected results shown in Table 5.

---

[11] Harold Gulliksen, Herbert S. Conrad and Norman Frederiksen, *Averages, Standard Deviations and Intercorrelations of Navy Aptitude Tests.* OSRD Report 1536. June 7, 1943. College Entrance Examination Board. Washington, D.C., Applied Psychology Panel, NDRC.

Herbert S. Conrad, *Item Analysis of Navy Aptitude Tests.* OSRD Report 3039. December 30, 1943. College Entrance Examination Board. Washington, D.C., Applied Psychology Panel, NDRC.

When the validity of Form 2 of the Basic Battery was studied, the training given in these schools had been improved. The result was a change in the relative validity of the various aptitude tests. The change is shown in Table 6 which gives the validities of the Form 2 tests, in comparison with the validities of the Form 1 tests, in these schools. For Form 2 one of the mechanical tests was the best predictor in each of the schools.

TABLE 6. Comparison of mean validity coefficients for the tests of the Basic Battery, Forms 1 and 2 (correlations corrected for curtailment in ranges of talent). For Form 2 the validity of the mechanical tests was relatively increased in these schools, all of which gave training in mechanical duties. The change was probably due to a change in school methods.

School		N	GCT	READ	ARI	MAT	MK(M)	MK(E)
Basic Engi-	Form 1	1480	.52	.52	.63	.52	.46	.39
neering	Form 2	1176	.58	.42	.50	.52	.62	.50
Electrical	Form 1	1747	.52	.52	.59	.44	.35	.49
	Form 2	1062	.48	.37	.49	.48	.36	.62
Gunner's	Form 1	1677	.38	.39	.31	.28	.40	.43
Mates	Form 2	809	.38	.32	.31	.50	.56	.54
Torpedo-	Form 1	880	.32	.35	.28	.27	.39	.35
men	Form 2	786	.29	.26	.24	.33	.54	.39

With the preliminary information at hand the Bureau and project developed and analyzed the Basic Battery.[12] Particu-

[12] The Panel contributions of general interest included:

H. Gulliksen, *Suggestions for the Revision of the United States Navy Mechanical Aptitude Test—Form I*. Project N-106, Memorandum No. 2. January 18, 1943. College Entrance Examination Board. Washington, D.C., Applied Psychology Panel, NDRC.

Herbert S. Conrad, *Statistical Analysis of the Mechanical Knowledge Test*. OSRD Report 3246. January 28, 1944. College Entrance Examination Board. Washington, D.C., Applied Psychology Panel, NDRC.

Norman Frederiksen, *Validity of an Experimental Battery of Aptitude Tests at the Ordnance and Gunnery Schools, Washington Navy Yard*. OSRD Report 3619. April 29, 1944. College Entrance Examination Board. Washington, D.C., Applied Psychology Panel, NDRC.

G. A. Satter, *Selection of Items for the U.S. Navy General Classification Test—Form 2 and the U.S. Navy Tests of Reading and Arithmetical Reason-*

lar attention was paid by the project to the internal characteristics of the tests. These are the characteristics which make a test a good measuring instrument. In this respect a test may be compared with a physical measuring instrument such as a voltmeter. A good voltmeter measures a significant range of electrical potential. It is relatively unaffected by disturbing chance variables like temperature or humidity. It is so built that it is hard for different men to make radically different readings for the same voltage. Its readings are relatively independent of unimportant characteristics of the circuit in which it is used. It measures voltage alone and its readings are independent of amperage or a-c frequency.

Aptitude tests, like voltmeters or thermometers, must be good measuring instruments. If they are, and if the aptitudes tested are significant (valid) for particular jobs, then the tests are successful. Among the more important internal characteristics of psychological measuring instruments are: (1) the difficulty of the tests and other attributes of the distribution of test scores, (2) test reliability, (3) test objectivity, (4) test independence, and (5) the homogeneity of the items of each particular test or subtest. These characteristics will be discussed with reference to the tests of the Basic Battery, Form 1.[13]

A test must be of a difficulty suited to the population tested. If the test is too hard or too easy one cannot differentiate between many of the men. In modern test practice one must differentiate not only the good men from the poor but within the

ing—*Form 2*. OSRD Report 3756. June 8, 1944. College Entrance Examination Board. Washington, D.C., Applied Psychology Panel, NDRC.

D. A. Peterson, *The Preparation of Norms for the Fleet Edition of the General Classification Test*. OSRD Report 4242B. October 10, 1944. College Entrance Examination Board. Washington, D.C., Applied Psychology Panel, NDRC.

Norman Frederiksen, *A Further Study of the Validity of the Arithmetical Computation Test*. OSRD Report 5302. July 3, 1945. College Entrance Examination Board. Washington, D.C., Applied Psychology Panel, NDRC.

[13] Herbert S. Conrad, *A Statistical Evaluation of the Basic Classification Test Battery (Form 1)*. OSRD Report 4636B. May 14, 1945. College Entrance Examination Board. Washington, D.C., Applied Psychology Panel, NDRC.

good group, within the average group, and within the poor group as well. To insure a suitable range of scores the difficulty of each item of a test must be determined and adjusted properly. It is wise to arrange items in order of their difficulty; otherwise the poor man may never have a chance to show what he knows. The purpose of an aptitude test is to discover a man's capabilities; it is not to prove him incapable.

The distribution of scores furnished by the various tests should be reasonably similar in shape so that the scores from different tests have comparable meanings. The distributions should be free of significant skewness. The shape of the distribution of scores should provide for the use of standard scores.

In its first form the Basic Battery was too difficult for the Navy population, but this defect was corrected in later forms. The shape of the distributions was satisfactory.

When a single individual is repeatedly tested, his score must be reliable, i.e. stable regardless of disturbing conditions. Chance fluctuations in score inevitably reduce the validity of a test. In part, reliability is a function of length and type of test. Shorter tests of a given type are generally less reliable than longer; low reliability is sometimes forced on a test by allowing too little time for it in an organized test program. Paper and pencil tests are generally more reliable than apparatus tests. Reliability is estimated by obtaining the correlation between a first and second administration of the test or between two random halves in a single administration. (Although the estimates derived by the two methods are really not comparable, in the case of the Basic Battery the values obtained by either method were similar in magnitude.) Reliability coefficients of the order of .90 are desirable and attainable for paper and pencil tests. Reliability coefficients of .70 or below suggest that a test receive special consideration with a view to improving its reliability. The tests of the Basic Battery, and the component subtests had reliabilities of the order shown in Table 7. The reliabilities shown for the Reading and

Arithmetical Reasoning Tests were a little lower than was desirable. The reliability of the Mechanical Aptitude Test was probably inflated since speed was an important determinant of score on this test. In general, the reliabilities were satisfactory when the length of time allowed to each test is considered.

TABLE 7. Reliabilities of the tests and subtests of the Basic Battery (Form 1). The figures are odd-even coefficients, corrected by Spearman-Brown. They are based on a random sample of 200 cases drawn from a national sample of 500.

General Classification Test	.94
Sentence Completion	.85
Opposites	.84
Analogies	.84
Reading	.85
Arithmetical Reasoning	.83
Mechanical Aptitude	.97
Block Counting	.95
Mechanical Comprehension	.84
Surface Development	.98
Mechanical Knowledge, Electrical	.89
Mechanical Knowledge, Mechanical	.90

Test scores must be objective; they must not vary with the individual who administers or scores the tests. There is no universal answer to the problem of making a test objective. In large-scale testing, as in the Army or Navy, tests will necessarily be administered by non-professional personnel of various backgrounds. Clearly the tests should be constructed so that the effects of variation in administration are reduced. On the other hand administrators should be trained adequately in proper methods of administration. No test can resist the effects of an administrator who shortens the time limits of the test because of an impending visit of high officers or who "gives everyone a chance" by an arbitrary increase in the time allowed. In the same way no test, even one which is machine-scored as practically all military tests must be, can be made independent of faulty clerical work in registering results.

Nevertheless a good test battery anticipates some of the above-mentioned difficulties. For example, tests made up of many very easy items give a score which depends on a man's speed. Minor variations of time limits then have a great effect on scores. If a test includes items which vary in difficulty and the items are arranged in order from easy to hard, it is usually less affected by arbitrary variations in time limits. The Basic Battery was relatively objective.

The scores on the various tests in a test battery should be independent of one another; a man obtaining a given score on one test should not necessarily obtain the same or similar score on another test. In part this is simply a way of saying that there is no value in measuring the same thing more often than is necessary to obtain a single, sufficiently precise estimate of its magnitude. But in part the problem is deeper; test independence is critical in the successful administration of military classification. If men with high scores on a test for one job necessarily have high scores on another test for another job, a reduction will occur in the number of satisfactory men available for assignment to either of the jobs. Taking a good man for one job eliminates him from consideration for the second, so that there may not be enough good men to supply all those needed in the two jobs. When manpower is limited and the number of essential jobs is great the problem is especially difficult. Low interrelations between valid tests simplify the problem.

The degree of test independence is expressed by the intercorrelations between scores from various tests. If test intercorrelations are low, the battery as a whole will provide for flexibility in assignment. Each job can receive a greater share of satisfactory men than if tests are not used at all. If test intercorrelations are high, appropriate assignment becomes difficult and classification degenerates into selection of a few generally good men for a few high-priority jobs or into the balancing of service units to insure that a number of generally good men are included in each. While these functions are important they do not compare in value with classification.

The problem of test independence is only partially solved for the military situation. In the Basic Battery, for instance, the various tests turned out to be correlated with one another to the extent of .40 to .70. The intercorrelations are shown in detail in Table 8, which reveals that the Arithmetical Reasoning Test was very closely related to the verbal tests, GCT, and Reading. The mechanical factor, as measured in the Basic Battery, was too closely related to the verbal factor to permit full use of differential prediction for various jobs. When the test intercorrelations were studied by the centroid factor analysis method of Thurstone, the results suggested that verbal and mechanical factors might be tested by the Battery, but not a separate numerical factor. The analysis also suggested the presence of a third factor, possibly a spatial factor.[14]

Although the Basic Battery only partially solved the problem, independent specialized tests were developed in World War II. The Radio Code Test—Speed of Response, to be described below in Chapter 5, is one of the best cases in point, since its correlation with the tests of the Basic Battery was very low.

In contrast to the preceding discussion the component subtests and items of a particular test should be homogeneous; they should all measure the same aptitude. Agreement among the components must be high enough to justify use of a single score for the test. Otherwise separate tests, or separate scoring systems, should be developed. The reasons for demanding homogeneity are simple and inherent in elementary mathematics; it is seldom useful to add a number of horses to a number of apples. The sum is usually less meaningful than either

[14] D. A. Peterson, *Factor Analysis of the New United States Navy Basic Classification Test Battery.* OSRD Report 3004. September 29, 1943. College Entrance Examination Board. Washington, D.C., Applied Psychology Panel, NDRC.

figure alone. In contrast it may be quite meaningful and helpful to add together the numbers of each of several kinds of
horses. The items of a test, or the component subtests, may
each measure a slightly different aspect of the aptitude in
question. The sum of the different items is then more meaningful, and more reliable, than any individual item. But if
each item measures a different aptitude the sum is necessarily
only a rather crude picture. The Basic Battery tests were
quite satisfactory in terms of homogeneity except in the case
of the Mechanical Aptitude Test. Reference to Table 8 will
show that the intercorrelations of the subtests were only .51,
.60, and .48 in this instance.

In addition to its research on the Basic Battery, the Panel
project on Navy aptitude tests developed and evaluated the
U.S. Navy Officer Qualification Test for use in the selection
of officer candidates. The project also conducted a number of
studies of tests and classification procedures. Some of these
will be described below.

In helping the Navy to modernize the basic tests of its clas-

LE 8. Intercorrelations among the tests and subtests of the Basic Battery (Form 1
ional sample, $N = 500$.

	GCT	Sent. Comp.	Opp.	Anal.	READ	ARI	MAT	Block Count.	Mech. Comp.	Surf. Devel.	MK(E)	MK(M)
T	—	.89	.90	.90	.81	.69	.60	.46	.57	.48	.53	.49
entence Completion		—	.74	.69	.73	.60	.55	.42	.48	.47	.46	.42
pposites			—	.71	.73	.61	.47	.38	.46	.36	.48	.46
nalogies				—	.74	.64	.59	.44	.59	.47	.50	.44
AD					—	.69	.56	.44	.52	.47	.51	.46
I						—	.61	.52	.51	.51	.47	.41
T							—	.87	.74	.86	.53	.55
lock Counting								—	.51	.60	.42	.45
echanical Comprehension									—	.48	.60	.62
urface Development										—	.36	.35
(E)											—	.78
(M)												—

sification system, the Panel fulfilled one of the responsibilities created at the time of its formation. The Navy's opinion of the success of the work of the Bureau of Naval Personnel and the Panel project on Navy aptitude tests was reflected in the eventual creation within the Bureau of Naval Personnel of a major subdivision of that Bureau, entitled Research. Under this subdivision the College Entrance Examination Board is continuing the contract work begun under the Applied Psychology Panel.

LEADERSHIP, INTEREST, AND EMOTIONAL STABILITY[15]

The studies of the general aptitude tests of the Navy Basic Classification Test Battery were supplemented by research on tests for combat leaders, for men's interests, and for emotional stability. In these supplementary projects Panel research was on a relatively small scale, although the problems were regarded as of very great importance. Research conditions and the nature of each problem were such that the odds were felt to be against successful results. Nevertheless the large possible gains were regarded as justifying preliminary studies. In the case of emotional stability a practical contribution to the winning of World War II was obtained in the form of a finished test, the Personal Inventory. In the cases of leadership and interest, Panel research did not go beyond the preliminary stages but did help to prepare the way for future practical contributions.

Leadership

The problem of leadership is of obvious importance to a country at war. Nevertheless it is not a problem which is easily subjected to psychological research. There is no satisfactory definition of the term leadership on which a quantitative approach can be based. The problem is particularly difficult if the emotional connotations are stressed, as in combat

[15] This section is based on *STR* I, Chapters 3, 4 and 5 by Dael Wolfle.

leadership. Thus when the Panel was requested[16] to study the selection of combat leaders, it agreed only to carry on a few preliminary explorations. Henry E. Garrett and Ernest M. Ligon made these studies.[17]

The original program called for a survey of information available in this country. If possible, the relation between leadership and existing information on aptitudes, prewar experience, and performance in Officer Candidate Schools was to be determined. The hope was that such a survey might yield information of sufficient interest to warrant a large-scale program of overseas research on a criterion measure for leadership. As the problem developed, a study was made of two classes of infantry officer candidates at Fort Benning, Georgia, and two classes of artillery officer candidates at Fort Sill, Oklahoma. Later, ratings of the leadership qualities of 176 overseas infantry company officers were obtained for the Panel by the Adjutant General's Office for comparison with school and other records.

At Fort Benning the two classes of officer candidates numbered about 200 men each. Some forty items of information obtained before or upon entrance to the Infantry Officer Candidate School were available for analysis. The criterion against which these items were evaluated was whether or not a candidate received his commission at the end of training. This criterion was mixed, including such unrelated material as school grades and the opinions of instructors and fellow students of "officer qualities" and "leadership potentiality." Some of the results are shown in Table 9. The criterion of receiving a commission was best predicted by the Army General Classification Test and next best by leadership as indi-

---

[16] The request originated in the National Research Council's Emergency Committee on Psychology and came to the Panel from the office of the Adjutant General of the Army.

[17] Under the contract with the National Academy of Sciences. The work is reported in Henry E. Garrett and Ernest M. Ligon, *Report to Applied Psychology Panel on Combat Leadership*, in Memorandum from W. S. Hunter, *Combat Leadership, Second Report on*. June 8, 1944. National Academy of Sciences. Washington, D.C., Applied Psychology Panel, NDRC.

cated by civilian occupation. A few other items such as father's occupation and amount of schooling seemed to warrant further study. One item which is commonly believed to be significant turned out not to be so in these data; vigorous athletic hobbies were no more favorable indications for a commission than mild sedentary hobbies. Participation in athletics had no significance as a possible selective item.

TABLE 9. Correlations of selected predictive items with the criterion of receiving a commission on graduation from the Infantry Officer Candidate School at Fort Benning, Ga. Items 1 to 3 are biserial and items 4 to 7 are tetrachoric correlations. Data are given for two classes of about 200 men each.

Predictive Item	Officer Class 304	Officer Class 311
1. Average of two administrations of Army GCT	.45	.46
2. Amount of schooling	.39	.07
3. Age at entrance to OCS	—.12	—.04
4. Father's occupation (leadership vs. non-leadership)	.12	.32
5. Athletic participation	—.08	—.10
6. Hobbies (vigorous and athletic vs. mild and sedentary)	.00	.02
7. Leadership as indicated by former civilian position	.20	.40

Since the Army General Classification Test was already used in the selection of officer candidates, the positive conclusion of the study was that more careful investigation of a candidate's previous leadership in civilian life might furnish significant data for prediction of success in securing a commission. If carefully developed and reasonably accurate, a system of scoring civilian leadership might be useful to the Army. Its usefulness would be fundamentally limited, however, as is the usefulness of all data on civilian occupations, by the fact that most Army personnel are derived from civilians under twenty-five years of age.

At the Fort Sill Field Artillery School, previous education, achievement in mathematics, and the Army General Classification Test were reasonably good predictors of commissioning. Selected correlations are shown in Table 10. In general, the results may best be interpreted as significant of the increased technical element in the performance of Field Artillery officers rather than of hope for the prediction of combat leadership.

TABLE 10. Correlations of selected items with the criterion of receiving a commission on graduation from the Artillery Officer Candidate School at Fort Sill, Okla. Data are given for two classes of about 200 men each.

Predictive Item	Officer Class 80	Officer Class 82
1. Army GCT	.29	.58
2. Education (number of years beyond elementary school)	.33	.36
3. Age	—.22	—.67
4. Participation in athletics	—.20	.16
5. Civilian occupation (rating from Barr scale)	.17	.32
6. Mathematical test (given in first week of school)	.48	.40

In a third study, ratings of efficiency in actual combat were obtained on 176 overseas company officers. These men were all graduates of the Infantry Officer Candidate School at Fort Benning. They were rated by regimental commanders and executive officers or by battalion commanders on a five-point scale of leadership in combat. There was a fair relationship between the combat ratings and ratings of the same men for leadership before they left Officer Candidate School. There was no relation to the Army General Classification Test, perhaps, however, because the group was a very select one in this respect. There was a tendency for the higher ratings to come from the age group 22 to 28. The most general conclusion of the study was that the schools were doing a fairly good job of eliminating men who would be unsuccessful overseas. The rating scale and the distribution of scores on it were as follows:

Rating	Number of Men	Per Cent of Whole Group Rated
Superior	23	13.1
Excellent	87	49.4
Very satisfactory	41	23.3
Satisfactory	17	9.7
Unsatisfactory	8	4.5

The scale and ratings are reproduced because they illustrate one major difficulty of research on leadership. According to the data, about 63 per cent of the group were superior or excellent leaders, and 86 per cent were more than satisfactory. Only 14 per cent were satisfactory or below in rating, and a mere 4.5 per cent were unsatisfactory. The group was said to be a fairly reasonable sample of officers. Such figures give reason to wonder whether there is any real problem of finding leaders. If only 8 men out of 176 are unsatisfactory there is little need to add psychological methods to those in regular Army use. Psychological methods could hardly reduce "failures" below 4 or 5 per cent.

Actually, of course, such figures are misleading. The term combat leadership may itself be prejudicial to a successful investigation. It connotes inspirational and emotional qualities which are probably rare or, at least, which very few people consistently exhibit. Except for these qualities combat leadership is probably based on quite different and less esoteric characteristics. An officer may be called good because he is a good administrator. He may know his subject matter. He may be able to train his subordinates. He may take care of the needs of his men. This list could be expanded to considerable length. Few men have all these qualities. Most officers probably have some of them. Thus, when the psychologist asks which are good officers, one man is named for one reason, another for a second, and still another for a third. When the question is phrased in terms of combat leadership, the situation is even more complex. In battle men may follow a leader because of his personality; or they may follow in spite of his personality, because of the quality of his noncommissioned officers, or because other officers have trained the men well. Until leadership is defined in a way which permits direct measurement of specific qualities, research on predictive items is unlikely to be useful.

It may be suggested that a job analysis and the development of achievement or proficiency measures are the next steps. They are needed not only for leadership qualities but for all officer qualities.

The job of an Army officer has never been carefully described in simple, factual terms. Merely to contemplate the need for a description is to suggest that the problem of officer selection may break down into a number of problems of the selection of a number of different kinds of officers. If common characteristics really exist for the various kinds of officers, or even for the successful officers of various kinds, these will become apparent in a job analysis.

When the significant elements in the jobs of the various kinds of officers are found, whether or not these be common to all officers, a second step in officer classification will be possible. A measure of proficiency or achievement in the significant elements can be developed. This is a relatively simple matter in so far as an officer's knowledge and skills are concerned and is already partly done. When performance of the significant parts of the job can be stated in quantitive terms, then it will be time for the third step, the development of aptitude and other classification tests for officers.

This is to propose a miscroscopic or analytic approach to leadership and "officer qualities." The global approach, tentatively tried by the Panel during the war, led to a recommendation that a detailed first-hand study of the officer in combat be made. Such a study would have provided at least the first step toward a job analysis of the officer in combat. Since the study was not made possible, the Panel believed that there was little point in continuing its work on combat leadership.

Interest: The Activity-Preference Test

A classification interviewer spends a considerable portion of his time in attempting to pin down the interests of the recruit. The recruit is asked whether he thinks he would like to be this, that, or the other kind of soldier or sailor. He is queried on his hobbies. He tells whether he enjoyed mathematics, languages, sciences, shop, or other subjects in school. His interests are probed because they are believed to be determinants of his future performance.

The same kind of information has been sought in civilian life by inventories and interest tests as well as by interview. Biographical inventories, now in successful use in particular military situations, include items on interest. An inventory or test differs from an interview in that an effort is made to standardize the information obtained on interests, to make the information relatively more reliable, and to state it in score form for ready use on a wide scale.

Since existing inventories and tests of interest were not suited for general military use, the Office of the Adjutant General of the Army requested the Panel to undertake a project on the development of an Activity-Preference Test for Classification of Service Personnel.[18] The research staff, who worked under T. L. Kelley, are shown in Table 11. Under this project Kelley developed and standardized the administration of an activity-preference test and worked out the scoring system. From the start of the project it was agreed that the

[18] A complete report is given in Truman L. Kelley, *Report on an Activity-Preference Test for the Classification of Service Personnel: Final Report of Project SOS-7.* OSRD Report 4484. December 21, 1944. Harvard University. Washington, D.C., Applied Psychology Panel, NDRC. Kelley's test is reprinted in the report.

TABLE 11. The research staff of the Panel project on a test of interests.

Contractor: Harvard University
Contractor's Technical Representative: T. L. Kelley
Staff: E. J. Fowler, H. M. Fowler, E. A. Haggard, M. J. Thwing

validation of the scores was to be undertaken by the Office of the Adjutant General.

The nature of the test is illustrated by the following samples from the 220 items which composed it:

1. Upon a free afternoon, which of the following would you like MOST and which would you like LEAST
   A. To go to a vaudeville show.
   B. To help organize a Boy Scout Troop.
   C. To take part in a play.
   D. To take the family out for a picnic.
2. Think of some situation in the past few years in which you have succeeded markedly. Which of these do you think was MOST and which do you think was LEAST responsible for the success
   A. I always worked hard to earn quick money.
   B. I had an excellent memory for people's faces.
   C. I was not timid in making constructive suggestions.
   D. I put more energy than most people into getting a job done.

The items are scored in such a way that each man comes out with a series of scores on each of several statistically-defined types of "interest."

The project turned over the test, a scoring table, and detailed statistical information to the Adjutant General at the end of 1944. Validation was not undertaken by the Office of the Adjutant General because of the pressure of other duties. The test remains unvalidated. It has the internal or statistical qualities of a good test for use with military personnel. It is believed to be promising, even though unvalidated, because great care was exercised to insure that the test would have the following characteristics:

1. The "right" answers are not easily guessed.
2. The alternative answers are genuinely competitive.

3. As nearly as possible all fields of life are represented.
4. No moral issues are involved in choices of activities.
5. Biographical information and attitudes toward oneself, as well as preferences for activities, enter into the score.

Emotional Stability: The Personal Inventory

Emotional stability is an obvious need of the soldier or sailor. It is a requirement not only in battle but also in training and even on leave from active duty. That our population contains a sizable fraction of mentally sick individuals is commonplace knowledge. It is often believed that abnormality is precipitated by life in military camps and by the obvious strain of battle. In a coordinated fighting unit even a single unstable individual can have serious and far-reaching effects. Thus one of the earliest requests to the Panel was to develop a test for emotional stability. This came from the Office of the Commander in Chief, U.S. Fleet and the Bureau of Naval Personnel and was supported by the Office of the Adjutant General. The project which resulted was named Research on a Personal Inventory and Other Tests for the Selection of Personnel for Especially Hazardous Duty.[19]

Division 7, the firecontrol division of the National Defense Research Committee, was already at work on the problem in cooperation with the Medical Research Laboratory of the U.S. Submarine Base, New London, Connecticut.[20] Under Division 7, C. H. Graham was investigating paper and pencil tests of the inventory type and apparatus tests of the psychomotor type as well as a test of speed of visual fusion under stress. The Panel requested Graham to take over its project

[19] A summary and complete bibliography of the work of the project are contained in W. C. Shipley and C. H. Graham, *Final Report in Summary of Research on the Personal Inventory and Other Tests*. OSRD Report 3963. August 1, 1944. Brown University. Washington, D.C., Applied Psychology Panel, NDRC.

[20] For Division 7, S. W. Fernberger had administrative responsibility for the research. Comdr. C. W. Shilling, USN (MC), represented the Submarine Base. Dr. Fernberger, Comdr. Shilling, and Lt. Comdr. W. A. Hunt, USNR, gave effective assistance and guidance to the Panel research in this field.

in the field and to concentrate his research on the development of the Personal Inventory. W. C. Shipley was directly responsible for the conduct of the work; the staff of the project is shown in Table 12.

TABLE 12. The research staff of the Panel project on the Personal Inventory.

Contractor: Brown University
Contractor's Technical Representative: C. H. Graham
Project Director: W. C. Shipley
Staff: R. N. Berry, H. R. Blackwell, F. E. Gray, D. B. Jones, H. J. Leavitt, F. A. Mote, N. Newbert, Eliot Stellar

The Personal Inventory was prepared on the basis of a detailed analysis of 100 psychiatric case histories drawn from the records of the Chelsea Naval Hospital. From the records 300 biographical items were formulated which seemed to differentiate the histories of these patients from the histories of average men. These were sifted in trial administrations to produce a long form of 145 items of which 60 were scored[21] and a short form of 20 items, all of which were scored. Included in the list of items were educational, occupational, social, and attitudinal items, as well as items describing symptomatic behavior. Some items referred to one type of abnormality, some to others. About an equal number of items for each psychiatric area were prepared. These took the form of such questions as the following in which the recruit was forced to choose between a symptomatic and a non-symptomatic alternative:

I have felt bad
more from head colds.

I have felt bad
more from dizziness.

21 W. C. Shipley, Florence E. Gray, and Nancy Newbert, *Item Analysis and Evaluation of the Scoring Stencil of the Personal Inventory.* OSRD Report 3315. February 14, 1944. Brown University. Washington, D.C., Applied Psychology Panel, NDRC.

This item form has both advantages and disadvantages when compared with simple questions, such as:

I frequently feel dizzy.   Yes.   No.

The advantages of the forced-choice question are that the undesirable or symptomatic behavior is less obvious than in the yes-no question, statistical treatment is simpler, and tendencies to answer "yes" or "no" to all questions are checked. On the other hand the forced-choice form has disadvantages; men object to being forced to choose between alternatives neither of which apply to them. Great care must be exercised to make the questions inoffensive.

One goal of research was to make the alternative answers equal in social desirability and inoffensiveness. The goal was never entirely reached. Perhaps it was for this reason that the test was not as satisfactory with officers as with enlisted men, in so far as Panel experience was concerned.[22] Certainly the administration of the Personal Inventory to a large number of young and vigorous Americans was never simple.

The Personal Inventory for enlisted men predicted whether or not a recruit would be discharged from the Navy for psychiatric reasons.[23] After the test had been standardized and the scoring system stabilized, it was administered to various groups of men; the results were filed and compared later with the psychiatrist's verdict on each man. In the studies reported here, care was exercised to insure that the psychiatrist was in ignorance of the test score at the time of the psychiatric examination. The problem of validation, in this instance, was

---

[22] One set of developments from the officers' form of the Personal Inventory will be described in one volume of the *Army Air Forces Aviation Psychology Program, Research Reports* (in press), U.S. Government Printing Office. According to a draft of this history, the Personal Inventory was regarded as one of the most hopeful tests for emotional factors in the research on the selection of Army pilots, bombardiers, and navigators.

[23] R. N. Berry, C. H. Graham, and F. A. Mote, *Results from the Long and Short Forms of the Personal Inventory and the General Classification Test.* OSRD Report 3962. July 31, 1944. Brown University. Washington, D.C., Applied Psychology Panel, NDRC.

whether the test would differentiate between recruits who were judged by the psychiatrist to be normal and recruits who were judged by the psychiatrist to be abnormal and so were discharged from the Navy. If a particular score on the test characterized a significant proportion of discharge cases and a negligible proportion of normal men, the score was differentiating. The score could then be used to predict the results of the psychiatric examination.

Figure 6 shows the best and poorest results obtained with

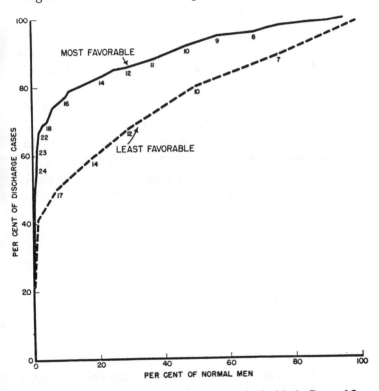

Figure 6. The most and least favorable results obtained with the Personal Inventory when this test was used as a psychiatric screen. The numbers on the points of the two curves indicate particular scores on the Personal Inventory. The position of a point on the abscissa indicates the per cent of all normal men who received that score or a poorer score on the test. The position of a point on the ordinate indicates the per cent of all men discharged for psychiatric reasons who received that score or a poorer score on the test.

the two forms of the Personal Inventory. Each point on the two curves represents a particular score on the test as shown by the numbers at the points in the figure. Poor (high) scores are at the left and the curves progress to good (low) scores on the right. The position of a point on the abscissa gives the cumulative per cent of normal men receiving the score. Position on the ordinate represents the cumulative per cent of discharge cases receiving the score. The curves show that poor scores on the test are received by a large proportion of the men who are discharged later. On the other hand, a very small proportion of the normal men receive poor scores on the test.

The meaning of the curves may be indicated more precisely by considering the point marked 17 on the least favorable curve. This point shows that a score of 17 or worse on the Personal Inventory was made by 50 per cent of all psychiatric discharge cases but only by 6 per cent of all normals. A score of 17 or worse on the Personal Inventory, then, suggested that the recruit would later be discharged from the Navy as a psychiatric casualty. Even the study which gave the poorest results for all forms of the Inventory showed it to be considerably better than the GCT as a test for emotional factors.

The results in Figure 6 are stated in terms of percentages, and in this case percentages give an impression of test efficiency which might be misleading. The actual number of men involved needs to be considered.

At the Newport station 40 of every 1,000 recruits were discharged by the psychiatrist. This means that the score of 17 or worse, which was used as the illustration above, identified 20 discharges and falsely identified 58 normal men. Suppose the test were used with a score of 17 as a cutting point at which men would be discharged from the Navy. It would then properly discharge 20 abnormal recruits but it would improperly discharge 58 normal men. In a tight manpower situation, this may be too great a sacrifice of normal men just to catch a group whom the psychiatrist calls abnormal.

On the other hand, the Personal Inventory can be used as a screening test for the psychiatrist. Continuing to use the same score from the least favorable curve as our illustration, the test may be expected to select a total of 78 men from each 1,000. These will include 50 per cent of all those who will later be discharged. The psychiatrist need examine only 7.8 per cent of all men in order to see half of all those whom he would ordinarily discharge. In so doing he cuts his examination time radically. Since only about 3,500 psychiatrists were available for all military and civilian needs in World War II, the time of the psychiatrist was at a premium.

The Personal Inventory was more successful as a psychiatric screen than any other test with which it was compared by the Panel or the Navy groups at Newport and New London. It was relatively independent of the General Classification Test ($r=.01$ to .35 in various samples). Its reliability was reasonably satisfactory even for the short form of 20 questions ($r=.66$ to .91 in various samples).

The closest competitor to the Personal Inventory as a psychiatric screen was the Cornell Selectee Index of the Committee on Medical Research, OSRD. Several questions from the Selectee Index were found to strengthen the Personal Inventory. The two tests were combined and placed in official use as a psychiatric screening test on January 24, 1945, by the Bureau of Naval Personnel in accordance with recommendations of June 29, 1944, from the Subcommittee on Psychiatry of the National Research Council's Division of Medical Sciences. The Panel concurred in the recommendations. The Personal Inventory had been in widespread unofficial use in the Navy, Coast Guard, and Maritime Service for well over a year before its official adoption.

The Panel made several efforts to obtain more direct evidence on the validity of the Personal Inventory for military duty. In one series of studies three kinds of stress situations were used. These included performance on escape-tower submarine training,[24] performance on a submarine sea patrol,[25] and performance in paratrooper training.[26] The first two gave little evidence of validity for the test but neither did they condemn it. The third gave positive evidence that the test is hopeful for prediction of performance under stress.

At the Army paratrooper training school at Fort Benning, 1,079 men were tested with the Personal Inventory before their training began. Of these, 778 successfully completed training and 301 failed. The Personal Inventory predicted all failures in paratrooper training with a validity coefficient of .39. It was superior to the Army General Classification Test for which the coefficient was .26.[27] There was no value in combining the two tests.

The Personal Inventory was about equally valid for all types of failure. The largest categories of failure were: insufficient desire to continue training, 89 men; refusal to jump from mock-up towers, 60 men; permanent physical disqualification, 34 men; and temporary injury, 67 men, some of whom would return to training later and some of whom would become permanent injury cases. The finding that the test predicted which men would suffer injury was not entirely unexpected.

In view of its validity in predicting failure in paratrooper training, it was recommended that the Personal Inventory be used in the selection of paratrooper candidates and that fur-

[24] C. H. Graham, F. A. Mote, and R. N. Berry, *The Relation of Selection Test Scores to Tank Escape Performance: Submarine School*. OSRD Report 3262. January 31, 1944. Brown University. Washington, D.C., Applied Psychology Panel, NDRC.

[25] G. A. Satter, *An Evaluation of the Personal Inventory and Certain Other Measures in the Prediction of Submarine Officers' Evaluations of Enlisted Men*. OSRD Report 5557. September 7, 1945. College Entrance Examination Board. Washington, D.C., Applied Psychology Panel, NDRC.

[26] G. A. Satter, *An Evaluation of the Personal Inventory for Predicting Success in Parachute School*. OSRD Report 4870. March 28, 1945. College Entrance Examination Board. Washington, D.C., Applied Psychology Panel, NDRC.

[27] Biserial *r* in both cases. The population was a normal group on the Army General Classification Test. The standard Navy scoring key for the Personal Inventory was used.

ther research on the subject be conducted. Neither recommendation was adopted by the Army.

In addition to the studies of stress situations in validation of the Personal Inventory, a follow-up study was made of the records of 1,446 sailors who had taken the Personal Inventory in recruit training.[28] All of these men had been accepted by the psychiatrist at the time of testing. One year after they had taken the test their service records were examined. Statistical analysis of the results showed that the Personal Inventory:

1. Differentiated between later discharges from the Navy and men active in the service after one year.
2. Differentiated between good and bad conduct cases.
3. Differentiated between those who became petty officers and those who did not become petty officers.

In similar comparisons the GCT was superior to the Personal Inventory in differentiating petty officers from non-petty officers but was inferior in the other two respects.

Although the Personal Inventory was thus shown to be successful in statistical predictions of behavior over a period of a year, the differences found were too small to warrant general use of the test in accepting or rejecting men for the service.

The Personal Inventory developed by Shipley and Graham was useful in saving the time of the psychiatrist. Studies of its other values point to the need for further research. It was demonstrated that factors leading to failure in stress situations are amenable to the test approach and that situations exist in training camps in which studies of emotional stability can be made. Since adequate studies of emotional stability in combat were not made in World War II, tests known to predict combat stability will probably not become available unless some future war makes such studies possible. From the military

[28] Walter C. Shipley, Florence E. Gray, and Nancy Newbert, *A Comparison of Personal Inventory Scores with Service Records One Year after Testing.* OSRD Report 3755. June 10, 1944. Brown University. Washington, D.C., Applied Psychology Panel, NDRC.

point of view, this makes it all the more important to follow up in situations like that of paratrooper training.

Emotional Stability: Battle Noise

In addition to its research on the Personal Inventory, the Panel evaluated the effectiveness of battle-noise equipment as a test for emotional stability.[29] A particular noise-reproducing system, which had been developed by Division 17 of the National Defense Research Committee as an extremely high-power, high-fidelity, sound-reproducing system, was used for these studies. Recordings made in battle, together with synthetic battle sounds, were played over the system. When men were exposed to the noise there were some who broke ranks and ran away. A few were taken directly to the psychiatric ward. Many more showed pallor, sweating, or other emotional signs.

For wide-scale use of the equipment as a classification device, direct observation of such symptomatic behavior was impractical. So men exposed to the noise were questioned, immediately following exposure, as to the symptoms experienced. Each man checked on a card whether or not he had experienced each of a number of symptoms described by the experimeter. The responses were filed and compared later with routine psychiatric diagnosis.

Since it was found that the reactions to battle noise agreed fairly well with psychiatric diagnosis, a control run was tried. Men who had never been exposed to battle noise, actual or synthetic, were given the questionnaire. They were asked to list the symptoms which they imagined they would feel if they heard the noise of battle. The imagined responses agreed with the psychiatric diagnosis to the same extent as did the re-

[29] The research was conducted under the general supervision of C. H. Graham. E. L. Hartley was in direct charge. Lt. Comdr. W. A. Hunt, USNR, and Lt. S. H. Britt, USNR, gave continual assistance and guidance. A summary and complete bibliography are given in Eugene L. Hartley and Dorothea B. Jones, *Final Summary of Research on the Use of Battle Noise Equipment.* OSRD Report 4931. April 12, 1945. Brown University. Washington, D.C., Applied Psychology Panel, NDRC.

sponses to the synthetic battle noise. Since no other criteria for validation of the battle-noise equipment could be obtained, the Panel recommended that the equipment be used to furnish greater realism in training. The recommendation was accepted.

The research of the Panel on leadership, interest, and emotional stability produced one test of practical military value and cleared the way for further studies of these topics. In the case of emotional stability there is reason to suppose that further practical values can be achieved at little cost.

THE PROCESS OF CLASSIFICATION

In addition to its research on general aptitude tests the Panel undertook many studies designed to evaluate or improve other aspects of classification.[30] A sample of a few will illustrate the character of the Panel's research and the nature of several basic problems in classification. These concern the validity of the classification process and aids to the interviewer.

It has been pointed out above that the job of the classification interviewer was difficult. It involved many decisions based on non-quantitative evidence. The question arose, therefore, whether the interview was done well enough to warrant the expense of the classification system, and it was desirable to validate the interview in order to have a basis against which to measure the effects of changes in the system.

For each recruit the interview resulted in an "Order of Assignment." The interviewer indicated the one school at which the recruit was most likely to be successful. The next most suitable school was then named, and third came a group of schools related in subject matter to the first choice. In any one school of the group the recruit should do reasonably well. Fourth came a group of schools related to the second most suitable school; if assigned to these, less could be expected of

---

[30] This section is based on *STR* I, Chapter 11 by John L. Kennedy and Chapter 12 by Norman Frederiksen.

him. In each of the four groups men were separated into three categories by quality: (1) exceptionally qualified on both aptitude and experience, (2) exceptionally qualified on aptitude but with no special occupational background, and (3) limited school possibilities. Finally there was the bottom group known as General Detail. The last included the men (about 40 per cent of all) for whom it was recommended that no special school training be given. In the operation of the classification system, assignment followed the order shown in Table 13. In one study of the validity of classification, the actual number assigned to a total of nine Navy schools for each category of the Order of Assignment was that shown at the right in Table 13. The study covered the period September 1943 to January 1944.

The men's performance in school was followed.[31] Of the

TABLE 13. The Order of Assignment of men to special training in the Navy (see text for explanation).

Order of Assignment	Number assigned from each category during study
12. 1st recommendation, Quality 1	3,383
11. 1st recommendation, Quality 2	10,036
10. 2nd recommendation, Quality 1	838
9. 2nd recommendation, Quality 2	2,890
8. 1st recommendation group, Quality 1	418
7. 1st recommendation group, Quality 2	1,372
6. 2nd recommendation group, Quality 1	36
5. 2nd recommendation group, Quality 2	115
4. 1st recommendation, Quality 3	10,374
3. 2nd recommendation, Quality 3	1,880
2. 1st recommendation group, Quality 3	1,280
1. 2nd recommendation group, Quality 3	127
0. General Detail	1,974
Total number	34,723

[31] G. A. Satter and Herbert S. Conrad, *Predicting Success in Service School from the Order of Assignment*. OSRD Report 5556. September 7, 1945. College Entrance Examination Board. Washington, D.C., Applied Psychology Panel, NDRC.

total group of 34,723 men, 3,231 were "rated," i.e. they were promoted to petty officer status because of their school performance and other characteristics. On the other end of the scale 2,501 men were failed in school; their special training was largely a waste of time and effort.

The potential success of the classification system can be indicated by calculating the proportion of men in each category of the Order of Assignment who were rated, and the proportion who failed the course. The results of the calculation are shown in Figure 7. In general the proportion of rated men from the top of the Order of Assignment is great and the proportion from the bottom is very small. The reverse holds for the failures. Figure 7 clearly indicates that the system predicted which men would succeed and which would fail in school. The classification procedures increased the flow of apt men to the schools and Fleet.

A major practical difficulty in the assignment of men is illustrated by the data shown in Table 13. The table shows that 1,974 men, or 5.7 per cent of the group studied, were assigned to school training despite the fact that the interviewer placed these men in the General Detail, special training *not* recommended. And 13,661 men, or 39 per cent of the group, were assigned to school although they were third, or poor, quality. It is evident that the manpower pool was inadequate to supply the need for men of higher quality.

Much might have been done that was not done to relieve the strain on the manpower pool. Some traditional service procedures needlessly increased the strain. For example, when a draft of men was required for a new ship or a beginning class at a school, the classification officer often received word to supply the men only a short time before they were needed. If he had the right kind of men at that particular moment, he supplied them. But if he did not have the right kind of men, he was given no time in which to accumulate them. Yet the dates of commissioning of ships or of starting new classes were known long before. The time required for finding the right men could have been made available. This and other similar practices prevented the full operation of a system for supplying the right men to the right place at the right time. It may be suggested that such a system is as badly needed as a system of

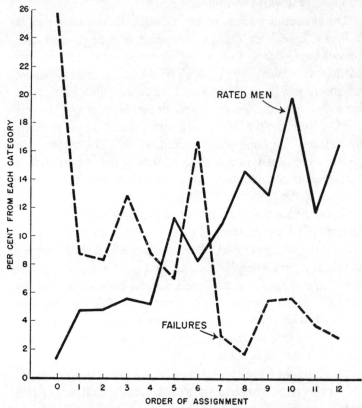

Figure 7. The success of the classification system in the assignment of men to Navy schools. Men whom the interviewer recommended for special training (high on Order of Assignment) included a large proportion who were rated, i.e., promoted to petty officer status. Men who were not recommended for special training (low on Order of Assignment) tended to fail their training course. See Table 13 for N's.

supplying the right materials to the right place at the right time in a modern mass-production factory.

The study of the validity of the interviewer's judgment was followed up by a comparison of the usefulness of quantitative and qualitative information in the interview.[32] The interviewer used aptitude test scores in arriving at his decisions. He also used purely qualitative information. Did the latter add anything to the former, subtract from the value of the former, or have no effect on the validity of prediction of school performance? It was almost universally assumed (implicitly, if not explicitly) by personnel men that qualitative information added to one's knowledge about a man and helped one to deal with him. The assumption was tested for one Navy situation.

The routine program of a particular Navy classification center in assigning 3,496 men to training school for Electrician's Mates furnished the data for the study. This center was chosen for the study on the basis of the belief of the Bureau of Naval Personnel that the interviewers were the best trained in the whole Navy classification system. A calculation was made of the correlation between the Order of Assignment furnished by the interviewers and grades in the school. The same type of calculation was made of the correlation between an aptitude test score, Mechanical Knowledge (Electrical) and school grades. The interviewer was supposed to weight this test score heavily in assigning men high on the Order of Assignment for Electrician's Mates.

If qualitative information added to the accuracy of prediction, the validity correlation for Order of Assignment would of course be higher than the validity correlation for the test alone. It turned out to be lower. Order of Assignment gave a correlation of .41; Mechanical Knowledge (Electrical) alone gave a correlation of .50. In this instance the classification process would have been improved by elimination of the interviewer and by assignment on a purely quantitative basis.

The Panel's problem was not merely to prove the greater significance of one or the other kind of information. It was to

[32] Herbert S. Conrad and G. A. Satter, *The Use of Test Scores and Quality-Classification Ratings in Predicting Success in Electrician's Mates School.* OSRD Report 5667. September 13, 1945. College Entrance Examination Board. Washington, D.C., Applied Psychology Panel, NDRC.

improve classification. On further analysis the same original data showed that qualitative sources of information probably might be made to help rather than to hinder the classification process. It was concluded that the interviewer tended to be too much impressed with the information which he himself collected in the course of the interview. If a recruit said that he had worked as an electrician in civilian life, or if the record showed that the recruit had served in the Navy as an apprentice, or "striker," Electrician's Mate, then a low aptitude score was given less consideration by the interviewer than it deserved. This tendency of the interviewer might be checked in two ways: first, by improving the training of interviewers, and second, by improving the accuracy of the interviewer's judgments at the time of the interview itself.

Aids to the Interviewer

Both methods of improving the interview were studied by R. K. Campbell with the assistance of the staff listed in Table 14.[33] The results[34] were incorporated in an experimental training course for interviewers which was turned over to the Bureau of Naval Personnel for comparison with the standard training course for interviewers. The end of the war prevented further research on the training of this type of personnel.

TABLE 14. The research staff of the Panel project on aids to classification.

Contractor: Stanford University
Contractor's Technical Representative: R. K. Campbell
Staff: E. A. Ingrahm, E. R. Keislar, M. M. Levin, A. C. Sherriffs, J. B. Sonderman

---

[33] A summary and complete bibliography of the work of this project are contained in Ronald K. Campbell, *Final Report in Summary of Research and Development of Classification Aids by NDRC Project N-116a.* OSRD Report 5647. September 11, 1945. Stanford University. Washington, D.C., Applied Psychology Panel, NDRC.

[34] Ronald K. Campbell, *A Study of Interviewing Techniques.* Informal Memorandum No. 3, Project N-116a. September 19, 1945. Stanford University. Washington, D.C., Applied Psychology Panel, NDRC.

The second approach was to simplify the judgmental processes required during the interview. An interview form was prepared on which were listed the specific elements considered important for success on a destroyer. For each element a weight was assigned. From these the interviewer could obtain a total point score for the man. A separate score had to be calculated for each job.[35]

The point-score method led to the invention of a simple form of calculating machine for the same purpose. E. R. Keislar developed an experimental model, which was named the Selectometer.[36] His objective was to relieve the interviewer of the necessity of remembering and weighting, more or less simultaneously, all the qualities which should be considered in the assignment of a man. In actual practice, interviewers were found to overlook some qualities, to be inconsistent in weighting, and, in fact, to be unaware in many instances of how they weighted the qualities which they did consider. The complexity of their task could be reduced by a simple calculating machine. A device was desired which the interviewer himself could operate at his desk and from which he could obtain an answer within a matter of seconds. The same calculations could be performed by the familiar card-sorting machines but these required extensive record sections and answers were necessarily long delayed. The Selectometer did the same thing on the spot and every interviewer could be provided with a Selectometer at a relatively small cost.

Keislar's first experimental model of the Selectometer provided a number of scales on which pointers were set to indicate scores on each of a number of significant characteristics. Aptitude test scores, height, weight, age, visual characteristics, etc., were taken directly from the records. Qualifications in sports, certain interests, and estimated ability as

[35] Max M. Levin, *A Point Score Method for Evaluating Naval Personnel.* OSRD Report 5197. June 12, 1945. Stanford University. Washington, D.C., Applied Psychology Panel, NDRC.

[36] Evan R. Keislar, *The Selectometer: A Classification Guide.* OSRD Report 4746. February 24, 1945. Stanford University. Washington, D.C., Applied Psychology Panel, NDRC.

a leader were also entered on the basis of the information collected in the interview.

In the first model of the Selectometer the calculations from these data were made by an optical method. For any single job the scale settings determined the density of filters placed in the path of a light beam. The intensity of the beam was measured by a photocell and meter which thus showed the sum of the scores on the various traits. For each job under consideration a separate light path and set of filters were required.

The optical model was soon replaced by a model of an electrical type of Selectometer. Resistors replaced optical filters with the principles remaining the same but with increased instrument reliability and with switching systems replacing the cumbersome photocell and numerous filters.

The development of the Selectometer was never carried beyond the pilot model stage. Sufficient use of the models in actual Navy classification was obtained to prove that the Selectometer can be made into a practical instrument. It has these interesting, potential characteristics:

1. It combines the advantages of multiple-cutting-score and multiple-correlation techniques. If a man has too high a score on GCT, for instance, he can be debarred from jobs where verbal aptitude is unnecessary and saved for other duties. On the other hand, a low GCT score can be permitted to temper the significance of a score on a mechanical test and the degree to which this is done can easily be varied throughout the scale of either trait. This is equivalent to providing all the advantages claimed for the "profile comparison method" but with the added advantage of standardized procedures.[37]

2. The distribution on any score may be shifted by any desired amount either in mean or standard deviation or both. Thus scores may be read in either absolute or relative terms. And priorities for the various jobs can be readily established and changed.

[37] The possible usefulness of the Selectometer principle in clinical psychology or clinical medicine may be suggested.

3. The Selectometer can provide for more certain use of the best available rating techniques in the case of those human characteristics for which tests and other measures are not available. There is considerable evidence from various sources to show that evaluations of people are more reliable when based on individual consideration of each of a number of fairly specific attributes than when global in nature. If it is necessary to use ratings, the interviewer can readily be required to make individual settings on the Selectometer for each characteristic.

To the research psychologist one of the most interesting features of the Selectometer was its revelation to the practical interviewer of just what kind of judgments went into the assignment of personnel to duty. The interviewer claimed that he considered a number of specific characteristics in evaluating a recruit. If he did so, the interviewer effectively assigned a score to each characteristic and weighted the score in relation to other scores. When classification interviewers were asked to set a pointer according to the recruit's potential leadership ability, for instance, the unsatisfactory and unreliable appraisal of this trait became apparent, and many interviewers balked altogether when they were asked just what weight should be assigned to this trait in the construction of the device. Thus, in those who used it, the Selectometer created a demand that fact be substituted for guesswork and precise conclusions for hazy ones. The need for reliable and valid measures in personnel work became clear.

These considerations, as well as those described in earlier sections of this chapter, suggest that the principle of the Selectometer is basic to the efficient use of manpower. The principle is to approach classification by the further refinement of quantitative methods. In the opinion of the Panel, the psychological problems of military assignment can best be solved by increasing the validity and precision of test and other measures on the one hand, and the precision of calculations based on the measures on the other hand.

CHAPTER 4

THE SELECTION AND TRAINING OF SPECIALISTS

IN the spring of 1942 the training of antiaircraft gunners was a major problem for the Navy. The need to protect ships from aircraft was emphasized by the course of the war in Europe, the events at Pearl Harbor, and naval engagements in the Pacific. Our ships, whether merchantmen, Coast Guard, or Navy, began to bristle with guns. Light and heavy machine guns, 3-, 4- and 5-inch guns appeared on the decks in astonishing quantities. To train men to operate these guns was a major undertaking. Ammunition was short and the production of guns, while immense, still provided relatively few for use in training. Barracks, class-rooms, training aids, and even courses of instruction had to be created, while the first recruits underwent such training as could be given. The recruits were unselected men, sent to gunnery school not because of any aptitude or experience with guns and similar equipment but because they happened to be available when a gunnery school had vacancies to be filled. Instructors were necessarily recruited hurriedly and en masse. As a result they were themselves somewhat unfamiliar with modern gunnery equipment and many were inexperienced as teachers of men.

The need for mass production of gunnery personnel was a major factor in the formation of the Committee on Service Personnel. Of the first two projects of the Committee, one was the Selection and Training of Naval Gun Crews.

Parallel difficulties were experienced by the Army and Navy with respect to other kinds of personnel than gunners: engineering and fireroom crews, amphibious personnel, night lookouts, radio code operators, radar operators, "telephone talkers," and other personnel using voice communication systems. In each case the Panel undertook a program of research on selection and training. In most instances the training situation was such that research on selection was necessarily slow and the opportunities were great for contributions to training. Thus it came about that these projects concentrated on training although in every instance some contribution was made to selection or classification.

In general terms the psychological approach to the training problem consisted of a survey of training methods and materials in use, followed by action programs designed to correct the deficiencies observed in the survey and by research designed to evaluate specific training methods and devices. The survey usually revealed that inexperienced instructors were failing to apply one or more of the following elementary principles of training:

1. Formulate the purpose of the job in terms that the trainee can understand.
2. Organize the material to be learned in simple, logical form, progressing from simple to complex, from concrete to abstract.
3. Secure the active participation of the trainee. Skills are not taught by words alone, nor are they taught merely by running a man through a job time after time.
4. In drilling men, insure that the right way of performing is practiced from the start. Decrease guidance as rapidly as the situation warrants.
5. Praise men whenever praise is reasonable and let criticism be rare.
6. Measure performance in a fair, objective, and standardized way. Motivate the men to competition and to self-competition. Keep accurate records of progress and make these available to the men. Knowledge of results is basic to the improvement of performance and it is more effective when it is given during or immediately after practice than when it is delayed.
7. If time may elapse between training and use of a skill, provide for original practice beyond the point at which men seem to cease to improve. Under the same circumstances provide for refresher training if possible.

8. Vary the discussion or drill whenever possible. Use visual aids, audio aids, models, synthetic trainers, and real equipment.

9. Distribute practice; do not concentrate it in long, fatiguing, or boring sessions.

In correcting any deficiencies in training the first step was to obtain a clear view of the nature of the job; a formal or informal job analysis was completed. In several instances the job analysis became the basis for a classification program. In others it provided the knowledge required to formulate outlines of courses, plans for specific lessons, and work books. With this work completed, several of the projects prepared courses for training the instructors themselves while others carried on an extensive advisory service to Army and Navy schools.

As opportunities and needs appeared these projects undertook research. They studied selection tests, developed and evaluated synthetic trainers, prepared and analyzed achievement and proficiency tests, and carried on experimental studies of training methods.

NAVAL GUN AND ENGINEERING CREWS

The program of this project was developed with special reference to the improvement of gunnery personnel by the application of the procedures in common use in industry.[1] As the methods of the project developed, the Navy requested their extension to the men who maintain and operate ship's engines, the engineering crews. The original program for gunners was formulated as:

Purpose: (1) To improve the program of instruction of men for gun crews so as to speed up the acquisition of knowledge and to make certain that men under training acquire the basic knowledge necessary for the effective performance of the tasks to which they are assigned; (2) to arrive at standard and more effective procedures for the use of training devices; (3) to develop and initiate instruction on "how to teach" for those who are responsible for training

[1] This section is based on STR II, Chapters 7, 14 and 15 by B. J. Covner.

personnel; (4) to prepare statements of qualifications and to develop and standardize tests for the classification of personnel with a view to the assignment of the most competent men to each of the gun crew stations.

Method: (1) A job analysis will be made of all the tasks involved in the effective operation of various types of guns. (2) A complete survey will be made of the material now available for instructing men in each such task. (3) Material now available will be supplemented by such other material as is needed and the entire content of instruction for each task will be surveyed with a view to rearrangement in the form of a curriculum calling for definite periods of instruction under specified conditions. This will be supplemented by job instruction sheets, standard examinations of progress, and such other aids as have been found effective in training industrial personnel. (4) By means of recording equipment a study will be made of the methods of instruction now employed and the interest of instructors will be enlisted in an improvement of their teaching methods. (5) There will be developed a short intensive course for training instructors on how to teach, following in general the program that has been so successful in training "job instructors" in industrial plants. While the initial phases of this program will involve the use of verbal material, consideration will be given to the preparation of films to be used in the later phases of the program. (6) An analysis will be made of the methods used in giving instruction on training devices, and appropriate step-by-step procedures will be outlined to produce the most effective results in using such training aids so that they will contribute most to performance on the combat tasks. (7) There will be included in this program a comparison of training devices to establish the relative usefulness of various instruments and to furnish clues which may be useful in the improvement of training devices. (8) It is expected that improvement in training methods will result in the development of objective criteria of performance on the task which can be used in the evaluation of selection methods. (9) When satisfactory criterion scores become available, aptitude tests and other procedures will be devised and experimentally analyzed with a view to providing more adequate methods for the classification of ship personnel.

The program, as outlined above, was carried through[2] in meticulous detail for both gun and engineering crews with the

[2] A complete summary and bibliography of the work of the gunnery project are given in M. S. Viteles, J. H. Gorsuch, and D. D. Wickens, *Memorandum on*

exception that objective criteria of performance were not developed, and in consequence, selection tests were not studied.

M. S. Viteles was in general charge of the work, with K. R. Smith and, later, J. H. Gorsuch serving as directors of the day-to-day field work. The complete personnel of the two projects is shown in Table 15.

TABLE 15. The research staff of the Panel projects on naval gun and engineering crews.

Contractor: University of Pennsylvania
Contractor's Technical Representative: M. S. Viteles
Project Directors: K. R. Smith, J. H. Gorsuch
Staff: Oscar Backstrom, Jr., A. G. Bayroff, F. K. Berrien, B. J. Covner, T. W. Forbes, P. H. Masoner, R. A. H. Mueller, Inez Murphy, W. E. Organist, M. H. Rogers, H. A. Voss, Leslie Watters, S. M. Wesley, D. D. Wickens, W. G. Willis
Subcontractor: New York University
Subcontractor's Technical Representative: D. B. Porter
Staff: Jacob Ain, Dora Albert, Oscar Reiss, S. G. Yulke

The projects began their work by analyzing the duties of the gunner or engineer. A job analysis is a systematic survey of the duties of personnel on a particular job. In making a job analysis of naval gunnery or engineering duties[8] information was collected by observing the performance of men on duty on shipboard. These observations were supplemented by discussions with specialist officers and instructors, by observing training, and by study of existing training, operational, and industrial literature. In the case of the 3″/50 caliber gun a

History and Final Report of Project N–105 (Final Report under Contract OEMsr–700). OSRD Report 6266. October 31, 1945. University of Pennsylvania. Washington, D.C., Applied Psychology Panel, NDRC. Similar material for the engineering crew project is given in M. H. Rogers, M. S. Viteles, and H. A. Voss, *Memorandum on History and Final Report of Project NR–106.* OSRD Report 6177. October 22, 1945. University of Pennsylvania. Washington, D.C., Applied Psychology Panel, NDRC.

[8] Morris S. Viteles and Kinsley R. Smith, *Job Analysis Procedure.* OSRD Report 1209. January 15, 1943. University of Pennsylvania. Washington, D.C., Applied Psychology Panel, NDRC.

time and motion study was completed.[4] The procedures gave the staff a broad view of the job under consideration. The information collected was formulated under the following headings:

Job title
Alternate title
Description of equipment used
Summary of duties
Supervision received or given
Relation to other jobs
Surroundings and conditions
Estimated time necessary to reach acceptable performance
Nature of training given
Stated qualifications for the job
Tools and materials used
Detailed description of duties: what was done, how it was done, and why.

The information collected in the job analysis served two purposes: (1) in order to help the classification interviewer the information was broken down into a statement of the psychological and other requirements of each job; (2) in order to help the inexperienced instructor the information was used as a basis for reorganizing the training courses and training the instructors themselves. The results of the program on training will be described. The work was done in close collaboration with naval officers. Members of the staff of the Bureau of Naval Personnel, gunnery or engineering instructors, and the Panel representatives worked together as a single team to improve training.

The job analysis defined the objective of the training courses. From the objective the curriculum was outlined. The outline was then developed into a series of lesson plans or guides for the instructor in teaching each specific unit of the

[4] By D. B. Porter and assistants on the subcontract with New York University.

course. The lesson plans varied from those which merely extended the course outline in slight degree to those which described in detail everything the instructor was to say and do. Typically the lesson plan was organized into five sections.

1. *Purpose.* This section stated the objectives of the lesson. These were made as specific to each job as possible.

2. *Preparation.* In this section the instructor was told what he must do beforehand in order to be properly prepared for the class. He was given lists of training aids, pictures, charts, and any other materials that should be prepared in advance.

3. *Presentation.* This section was the principal part of any lesson primarily devoted to giving information to the trainees. If the lesson was primarily devoted to drill, the section on presentation served merely as an introduction to the drill period.

Questions such as the following were considered in planning the presentation section of a lesson plan:

a. What type of introduction should be provided?

b. What technical terms should the trainees be expected to learn?

c. In what order and how should these technical terms be taught?

d. How much straight lecture material should be used?

e. At what points in the presentation can training aids be used to add interest or to make things more understandable?

f. How will those aids be used?

g. Where can questions add to the effectiveness of the presentation?

h. What type of questions should be used—those calling for one-word answers or those demanding explanations of general principles? Can thought questions which require the trainee to tie together parts of the presentation be used to advantage? It is believed to be sound practice to provide actual questions and their answers in the printed lesson plan. Such questions are probably better than those made up on the spur of the moment. The classroom situation should be allowed to determine which ones will be used, but supplying the instructor with a good set of questions is frequently a valuable service to him.

i. What allowances should be made for differences in the background of groups taught? For example, if a particular group is already acquainted with the subject matter, which sections of the material should be omitted and what should be substituted?

j. How much time should be given to a review of the last lesson, to a preview of this one, to showing the relation between this lesson and other parts of the course?

4. *Practice.* In this section of the lesson plan provision was made for actual practice or drill. Questions such as the following were considered in planning practice or drill periods.

a. How much time should be allowed for practice with the various methods suggested?

b. How can variety, novelty, or competition be used to maintain interest during a long drill period?

c. How frequently should trainees be shifted from one activity to another? Should rest periods be used?

d. What degree of skill should be required of trainees?

5. *Review and Summary.* Whenever possible, the lesson plan provided for a review or summary of material covered in that lesson, and also some material covered in earlier lessons. This section should include:

a. A review of the major points covered.

b. An evaluation of the progress of the group. What were the major difficulties encountered? What parts of the material need further study, or what procedures need further practice?

c. A brief statement of the next lesson, showing how it is related to the present one.

In accordance with this general program, job analyses and

lesson plans were prepared for the 20 mm. and 40 mm. machine guns, the 3″/50, 4″/50, 5″/38 caliber guns and the main battery, as well as for fireroom and distilling plant operation, and for other duties.[5]

The program included the development and experimental analysis of training aids and synthetic trainers and the preparation of drill manuals and lesson plans to insure satisfactory use of such devices.[6] Some of the project's contributions to training aids are described in Chapter 7.

In training gun and engineering personnel the culminating step was to prepare courses for the instructors themselves.[7] Each course was accompanied by a kit of training aids and by phonograph records of sample lectures, explanations, and drills. The recordings illustrated good and bad teaching techniques with specific reference to the particular kind of training covered by the course. The instructors were encouraged to evaluate their instruction by listening to voice recordings of their own lectures and demonstrations.

The work of the projects on gun and engineering crews was thoroughly integrated into Navy practice. The Bureau of Naval Personnel, which carried on the same type of program

[5] K. R. Smith and J. H. Gorsuch were assisted in the preparation of these materials by B. J. Covner, P. H. Masoner, R. A. H. Mueller, W. E. Organist, L. R. Watters, S. M. Wesley, and others. Samples of the instructional material developed for naval gun crews are described in M.S. Viteles *et al., Unit Lesson Plans for 4-Day Course in 20 MM Gunnery*. OSRD Report 1371. May 1, 1943. *The 40 MM Gun: Training and Indoctrination of Crew, Instructor's Handbook*. OSRD Report 1591. July 1, 1943. University of Pennsylvania. Washington, D.C., Applied Psychology Panel, NDRC.

Typical materials developed for engineering crews are described in Bernard J. Covner, John H. Gorsuch, and Morris S. Viteles, *Memorandum on Detailed Fireroom Operating Procedures for Destroyer Escort Vessels, Turbo-Electric and Turbo-Geared Types*. OSRD Report 3444. March 15, 1944. University of Pennsylvania. Washington, D.C., Applied Psychology Panel, NDRC.

[6] D. D. Wickens and A. G. Bayroff, under the immediate direction of J. H. Gorsuch, were primarily responsible.

[7] Kinsley R. Smith *et al., Memorandum on How to Teach Gunnery*. OSRD Report 3876. July 11, 1944. Morris S. Viteles and John H. Gorsuch, *How to Teach Engineering: A Short Course in Effective Teaching Methods for Engineering Instructors*. Project NR–106, Memorandum No. 4. October 9, 1944. University of Pennsylvania. Washington, D.C., Applied Psychology Panel, NDRC.

for an extensive series of Navy jobs, secured wide use of the manuals in many training stations and ships. The net effect was a systematic improvement in Navy training practices.

AMPHIBIOUS PERSONNEL

The Navy's amphibious training program was a new development necessitated by the conditions under which World War II was fought.[8] The craft used in amphibious warfare were under continuous development. Crews were small; control of personnel had to be decentralized. Officers were inexperienced; the men were chiefly raw recruits. In the winter of 1944 a former member of the Committee on Service Personnel, Capt. P. E. McDowell, was transferred to duty as Training Officer on the staff of the Commander, Amphibious Training Command, U.S. Atlantic Fleet. The Panel soon received a request for research on the Selection and Training of Amphibious Personnel.

The program which resulted[9] was a joint activity of the Bureau of Naval Personnel, the Amphibious Training Command of the Atlantic Fleet, and the Panel. For the Panel K. R. Smith was in charge. The project staff is shown in Table 16.

TABLE 16. The research staff of the Panel project on amphibious personnel.

Contractor: Pennsylvania State College
Contractor's Technical Representative: K. R. Smith
Project Director: H. A. Voss
Staff: T. L. Bransford, L. S. S. Hoffman, G. E. Lindzey, E. T. McDonald, D. J. Moffie, H. C. Reppert

---

[8] This section is based on *STR* I, Chapter 11 by J. L. Kennedy and *STR* II, Chapter 9 by Dael Wolfle.

[9] A summary and complete bibliography are given in Kinsley R. Smith, *Final Report in Summary of Work on the Job Analysis, Qualification and Placement of Personnel in the Amphibious Force.* OSRD Report 5422. August 8, 1945. Pennsylvania State College. Washington, D.C., Applied Psychology Panel, NDRC.

Two types of activity were conducted. The first was the development of a classification program and an attempt to validate tests and interview ratings against the shipboard performance of the men classified. The second was a training program having a marked similarity to that described above for gun and engineering crews.

The classification program for amphibious personnel adapted the standard tests and procedures of the Navy to the special conditions of the amphibious training program. It provided not only for improved use of individual men but also for "balancing" the crews of the whole amphibious fleet. Balancing crews by aptitude was a necessity since in a coordinated attack by many small craft a few inept crews could create a disaster. When the Panel began work it discovered that an enterprising officer was able to round up a crew which consisted entirely of the best available men. This meant that the next crew was drawn from a pool with the top quality removed. The process continued and resulted in marked unevenness of crews. The problem was to balance whole crews against one another and to secure the most effective assignment within single crews. The result of the classification program was a Manual of Classification Procedures for Amphibious Training Bases.[10]

The training program included job analysis of all billets in attack boats and the landing ships (tanks, medium, and infantry).[11] Special training manuals were prepared for piloting and poison gas defense. The staff contributed heavily to a general Training Manual[12] of the Amphibious Training Command.

It is difficult to obtain a scientific estimate of the value of the classification and training procedures developed for the

[10] From the Amphibious Training Command, United States Atlantic Fleet, Naval Operating Base, Norfolk, Va. January 9, 1944.

[11] E. T. McDonald, G. E. Lindzey, and D. J. Moffie were especially concerned with this phase of the project's work.

[12] From the Amphibious Training Command, U.S. Atlantic Fleet, Naval Operating Base, Norfolk, Va. September 1, 1944.

Amphibious Training Command. For classification a large-scale validation program was undertaken by the Panel, but it was necessarily dependent for a criterion on rather general ratings from many officers. Each officer was familiar with a few men only. The resulting validity coefficients were low.[13]

The practical phases of the work of the Navy and Panel groups were highly regarded in the Navy. Captain McDowell was cited, and on February 16, 1945, the Commander-in-Chief, U.S. Fleet, directed the Chief of Naval Personnel to develop for fleet-wide application a classification program similar to that used in the Amphibious Training Command of the Atlantic Fleet.[14]

CARGO HANDLERS

In the winter of 1944, as shipping problems rose to their height in the Pacific, the Panel was asked by the Commander, Operational Training Command, U.S. Pacific Fleet, to study the selection and training of cargo handlers for Assault Personnel Auxiliary and Assault Cargo Auxiliary ships. A large number of these ships were under construction and very few civilian winch operators were available, and inexperienced cargo handlers were inadvertently destroying much high-priority cargo. F. L. Ruch and the staff shown in Table 17 studied the problem.[15]

Preliminary studies were made of a large number of selec-

[13] Kinsley R. Smith et al., The Effectiveness of Classification Data in Predicting Billet Performance in Training in the Amphibious Force. OSRD Report 5198. June 12, 1945. Pennsylvania State College. Washington, D.C., Applied Psychology Panel, NDRC. See also D. R. Miller, Final Report in Summary of Work on the Selection of L.C.V.P. Coxswains. OSRD Report No. 5203. June 14, 1945. National Academy of Sciences. Washington, D.C., Applied Psychology Panel, NDRC.

[14] By second endorsement to ComPhibTraLant ltr. FE 25/A7–3, Serial 0043, January 24, 1945.

[15] A summary and complete bibliography are given in Floyd L. Ruch, Final Report in Summary of Work on the Selection and Training of Cargo Handling Teams for Combat Laden Vessels. OSRD Report No. 5140. May 29, 1945. University of Southern California. Washington, D.C., Applied Psychology Panel, NDRC.

tion tests. The results were turned over to the Bureau of Naval Personnel, which followed up with its own research.

In the field of training, F. L. Ruch and A. H. Nace developed a synthetic electric winch trainer. The device was a miniature winch in so far as size, speed, and lifting power were concerned. The controls, however, had the same dimensions

TABLE 17. The research staff of the Panel project on cargo handlers.

Contractor: University of Southern California
Contractor's Technical Representative: F. L. Ruch
Project Director: Anthony Nace
Staff: Arthur de Golyer, Willard Lusk, Daniel Miller

---

as the winches then being installed on the cargo vessels. Following trial use, a set of six lectures and practice drills were developed. These were incorporated in a manual with a proficiency test, a rating scale, and instructions for construction and maintenance of the trainer. Ruch made several efforts to validate the trainer, but crowded schedules prevented completion of all but one study, which suggested that the trainer was valid. The device was placed in use in three operational training stations on the west coast.

NIGHT LOOKOUTS

World War II saw the first emergence of systematic large-scale night warfare. Basically, of course, night operations were made possible by the development of radar and other "blind fighting" equipment, but for many purposes the human eye was superior to radar and other substitutes.[16] At all times when human judgment was concerned in an operation, efficiency was improved if objects could be seen with the eyes. The speed of operation was increased by the use of the eyes, the number of men and the need for communications between

[16] This section is based on *STR* I, Chapter 9 by W. E. Kappauf and *STR* II, Chapter 6 by Dael Wolfle.

them were reduced, and the soundness of interpretation was improved.

For many years it has been well known to laboratory scientists that there are differences between the use of the eyes in daytime and their use at night with no artificial light. At night colors cannot be distinguished. Stereoscopic depth perception is lost completely. All visual functions are impaired after the sun sets. At a low level of illumination small objects become invisible altogether if a man looks directly at them; in order to see them he must look a little to the side, above, or below, counteracting the normal habit of looking directly at the object of interest. And staring must be avoided altogether, since objects fade out in a few seconds of fixation of the eyes on one point. Finally, the eyes at night rapidly lose and slowly regain efficiency each time that they are exposed to light.[17]

These and other facts which relate to the differences between the rods and cones of the retina of the eye were applied by the British to the problem of stopping German bombers in the Battle for Britain. Radar guided the night fighter to the general vicinity of the enemy plane; the eyes were needed to shoot the enemy down. British night fighter pilots were screened to eliminate the night blind; they were trained to use the "corner of the eye" in night search procedures; their ready rooms were darkened; their equipment and operating procedures were modified to eliminate sources of glare in the field of view; and other changes were made to secure with certainty the most effective use of the eyes at night.

In this country one of the major sources of interest in military psychology developed from the British use of the facts about night vision. The differences between the ordinary use of the eyes in daylight or in artificial light and their use under

[17] A pamphlet designed to inform officers of the facts about night vision and to stimulate practice in the use of the eyes at night was prepared by Marjorie Van de Water of Science Service and the present writer, *The Best Way to Use Your Eyes at Night*. Science Service. August, 1942. The pamphlet was reprinted in the Infantry Journal and a number of other publications. After revision it was published separately by the Armed Forces Institute and the Bureau of Naval Personnel.

complete blackout conditions were a relatively old story to the laboratory scientist. The differences came as a dramatic surprise to the average man. Many highly-placed English politicians and military men believed that German night bombing was stopped not only by radar but also by the effective use of the principles of night vision. In 1941 and 1942 the British attitude led to the development in this country of selection tests for the "night blind," to new night operating procedures, to the use of red goggles as an aid to dark adaptation, and to the substitution of red light for blue as the standard artificial light in our Navy.[18]

Scientific effort greatly improved American night operations, but three psychological problems remained in 1943: (1) selection tests for night vision were many in number but little was known of their validity and other characteristics; (2) the ability of personnel to operate by eye at night was only vaguely defined; and (3) there was still no commonly used, organized, training system on how to use the eyes at night. The Navy presented these three problems to the Panel in a request for research on the Selection and Training of Night Lookouts. The Panel secured the services of C. H. Wedell and the staff shown in Table 18 to carry out the desired studies.[19] Capt. C. W. Shilling, USN (MC), and Lt. W. S. Verplanck, USNR, participated in the work, much of which was done at the Medical Research Laboratory, U.S. Submarine Base, New London, Conn.

[18] The OSRD Committee on Medical Research, Division 16 of NDRC and the Medical Research Division of NRC contributed heavily to these developments. The NRC Committee on Human Aspects of Observational Procedures was active in liaison. Later in the war the liaison function was assumed for all visual problems by the Joint Army–Navy–OSRD Vision Committee, whose Executive Secretary, D. B. Marquis, was attached (for administrative purposes) to the Applied Psychology Panel as a Technical Aide. The joint committee is continuing in the post-war period as the Joint Army–Navy–NRC Vision Committee.

[19] A summary and complete bibliography of the research of the project are given in C. H. Wedell, *Final Report in Summary of Work on the Selection and Training of Night Lookouts*. OSRD Report No. 4342. November 15, 1944. Princeton University. Washington, D.C., Applied Psychology Panel, NDRC.

TABLE 18. The research staff of the Panel project on night lookouts.

Contractor: Princeton University
Contractor's Technical Representative: H. S. Langfeld
Project Director: C. H. Wedell
Staff: A. C. Hoffman, L. H. Lanier, P. C. Miller, W. C. H. Prentice,
  Sidney Sanderson, R. C. Young

The Panel had had many previous opportunities to engage in experimental studies of night vision. But none of these had provided for research on night vision under practical operating conditions. Other organizations had the same experience; research was desired but the conditions offered for research were such that only abstract conclusions could be reached. From the psychological point of view, the early work had concentrated too heavily on the visual aspects of night operations at the expense of other phases of equal importance. In developing the request for the Panel project, Lt. S. H. Britt, USNR, and Capt. J. H. Thach, Jr., USN, of the Readiness Division, Office of the Commander-in-Chief, U.S. Fleet, specified that the research not only should but could be carried out under relatively realistic conditions which would permit a study of the whole work of the night lookout.

The first problem was to undertake an evaluation of the numerous devices developed to test the night vision of military personnel.[20] For the most part these represented applications of one or another laboratory psychological method to a miniature night vision situation. One test, the Radium Plaque Adaptometer, was already on order for general Navy use. It presented a black T-shaped figure in various positions on a five-degree field of low illumination provided by a radium-coated plaque. The subject's task was to identify the position of the T. It was hoped to study the validity of this and other tests, not as abstract tests of night visual capacity but as

[20] Carl H. Wedell, *A Study of the Prediction of Night Lookout Performance.* OSRD Report 3357. March 15, 1944. Princeton University. Washington, D.C., Applied Psychology Panel, NDRC.

predictors of performance in the visual duties of the night lookout.

A total of nine tests of night vision, including the Radium Plaque Adaptometer, the Royal Canadian Navy Adaptometer developed by Hecht, and the NDRC Model III Adaptometer[21] were administered to 150 inexperienced seamen at New London. The seamen were then trained for three days on a course of instruction for night lookouts. A retest with all nine tests followed. Then the men were taken aboard ship for several nights where they served as lookouts against a target vessel.[22] The target vessel made repeated approach runs. Over a telephone each man reported the bearing of the target when he first saw it on any run and repeated his report at stated intervals thereafter. The range and bearing of the target were continuously observed by radar.

For a number of reasons the experiment provided little direct information on the validity of the night vision tests. Most significant among the reasons was the unreliability of the criterion measure and all the test scores. The odd-even reliability coefficient for performance on shipboard was only .56. And for the four tests whose reliability could properly be estimated by the correlation method, the test-retest reliability coefficient ranged from .19 to .53. None of the remaining tests gave indication of reasonable reliability.[23]

The conclusion was that no one of the nine tests was satisfactory. The conclusion was confirmed for the Radium Plaque Adaptometer in a follow-up study made by the project as a part of its further investigations of the night lookout (see below). Measurements were made of the performance of 63 night lookouts on a convoy leader at sea. Of the 63 men, 50 had previously passed the Radium Plaque Adaptometer Test

---

[21] The NDRC Model III Adaptometer was developed in accordance with general specifications provided by the writer.

[22] In the shipboard studies experimental difficulties reduced the population for the validity study to a total of 58. The major difficulty was the unexpected use of the area by airplanes from another base for searchlight practice.

[23] On analysis by *chi*-square techniques.

and 13 had failed. The average performance of those who had failed the test was better than the average performance of those who had passed the test.

The use of the Radium Plaque Adaptometer on all men of the Navy would have required the expenditure of several million man-hours in the time of examiners and examinees. Although experimental studies in the Army Air Forces and Army Ground Forces suggested that similar tests might have validity for night operations, the Panel recommended that the Radium Plaque Adaptometer Test not be used in the Navy. The Bureau of Naval Personnel accepted the recommendation and directed that Radium Plaque Adaptometer scores "not be used in determining qualifications for any school or rate. . . . Scores for this test should not be recorded on the Enlisted Personnel Qualification Card."[24]

The second phase of the project's work followed the Navy's request to make an analysis of the duties and successes of officer and enlisted personnel in various night lookout assignments with reference to the specific abilities required for success. The analysis was made aboard a cruiser which was serving as the leader of a convoy. C. H. Wedell and W. C. H. Prentice observed and collected data on the lookouts of the cruiser during one round trip to Europe.

The lookouts traditionally have played an important role in naval operations. They watch for other ships, friendly or enemy, or for lights, smoke, debris and water patterns that may indicate the presence of other ships. They help in navigation as well as in combat. In more recent years, submarines, periscopes, torpedoes, and airplanes have been added to the list of things to detect and report.

In one sense the task of the lookout is a very simple one. He merely scans his sector of the sea or sky and reports what he sees. But in another sense his difficulties are fairly great. Look-

[24] *Manual of Procedures for U.S. Naval Classification Centers*, p. 150, NavPers 15082. February 1945. Washington, D.C., Navy Department, Bureau of Naval Personnel.

ing steadily at a sector of the sea with nothing in sight but the sea itself is not conducive to alertness. Questions of fatigue and of motivation must be added to the more obvious visual aspects of the lookout's job. Airplanes approach with such high speed that systematic and rapid scanning becomes necessary. At night the lookout must learn to acquire and maintain dark adaptation, and he must learn how to use his eyes most effectively in the dark if he is to pick up targets in time for appropriate action to be taken. He must learn a number of special techniques in order to make accurate estimates or readings of the range, elevation angle, target angle, and relative or absolute bearing of a target. He must learn how to adjust and use binoculars and other equipment in order to increase the range and accuracy of vision. The non-sensory aspects of the task are at least as significant as the sensory aspects.

Some of the non-sensory aspects of the task of the lookout were reviewed by the project in a report which stressed the importance of the organization of the ship's crew as it affected lookouts, the low prestige of the lookout in the existing organization, the training of the lookout and the lookout officer (which concentrated heavily on aircraft and ship recognition to the exclusion of other parts of the task), and the need for periodic inspections of lookouts on duty.

The third phase of the project's work concerned the measurement of the performance of night lookouts at sea. It is important to the services to know what they can expect of their men in the use of the eyes at night. In the case of the night lookout the ability to locate other ships is one of several determining factors, for example, in the spacing of ships in convoys, in the number of protecting ships provided to convoys, and in the number and location of lookouts on board each ship. To some extent the records of routine operations provide empirical answers to many of these questions. Direct measurement of lookout performance should supplement the analysis of operational data. By direct measurement the evidence can be made more precise at the expense of some loss

PLATE 1. A training device for the night lookout. When the dark-adapted eye is placed in the eye-piece of the box, one sees a set of the interchangeable figures shown in Plate 2. Each set of figures is printed on a separate curved plastic plate. Illumination is provided by a similar plate, coated with a radium salt and permanently mounted behind the stimulus plate. The device can be used in a semi-darkened room.

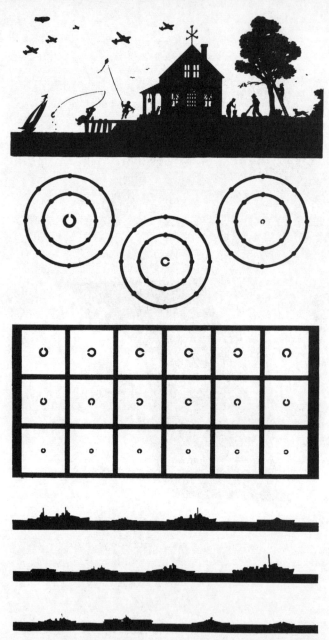

PLATE 2. Interchangeable stimulus cards for the night lookout trainer (see Plate 1). The course of dark adaptation can be followed by studying the view of the house in which smaller and smaller objects become visible as dark adaptation proceeds. The rings illustrate the value of night-time use of the "corner of the eye." The C-figures are used as a rough test of ability to see at night, and the ship silhouettes show the appearance of various types of ship on the horizon at night.

of realism. Among many English and American contributions to such measurement was that of Wedell and Prentice,[25] who sought answers to these three questions:

1. What can the average lookout see under various night conditions?
2. How much difference exists between a good night lookout and a poor one?
3. How consistent are lookouts?

On the convoy leader, studies were made of the ability of the ship's regularly assigned lookouts to spot targets visually at night. The lookouts had been chosen for their duties by their respective division officers and were under the supervision of a Lookout Officer and an Assistant Lookout Officer for purposes of organization and training. None of the men were petty officers. There was reason to believe that these men were typical of those who generally do lookout duty in the Fleet.

Observations were made only during the hours of darkness. One of the investigators stood on the open deck at the level of the forward sky lookouts. He made frequent qualitative and quantitative records of weather conditions and visibility. The second investigator was stationed at the ship's radar gear. Both were in direct communication with all lookouts and with each other on the regular lookout circuit of the ship's telephone system.

The investigator at the radar gear phoned each lookout in turn, asked for the bearing of the most distant ship the lookout could see in his sector, recorded the bearing of the target as reported by the lookout, and obtained the range and bearing of that target from the duty radar operator. At the same time the investigator recorded the name and station of the lookout, the date, and the time. Weather, sky, and sea brightness conditions were obtained by phone from the investigator

[25] W. C. H. Prentice, *A Study of the Performance of Night Lookouts Aboard Ship*. OSRD Report 4087. October 15, 1944. Princeton University. Washington, D.C., Applied Psychology Panel, NDRC.

on the open deck and were also recorded. The type of ship sighted (transport, tanker, destroyer escort, etc.) was obtained later from a chart of the convoy.

At the beginning of each watch, men came directly from compartments lighted only by a dim red light. Testing did not begin until the men had been on watch for fifteen minutes or more, thus all men were dark adapted before being asked to make any report. All tests were made when the lookout was using standard Navy 7 × 50 binoculars. With few exceptions all bearings were read from alidades. When the alidade could not be read, the bearing was estimated by the lookout. On most nights it was possible to secure reports during only one watch. Moonlight and the short hours of summer darkness prevented a longer observation period.

The raw data from these observations consisted of bearings and ranges of the most distant ships visible to lookouts when they were asked for a report. In their original form, these data provided no means of comparing lookouts directly. The size and distance of the target ship, the angle of the target, the illumination, and the height of the lookout all varied. It was therefore necessary to introduce a considerable number of corrections to make the reports comparable. Two main types of correction were necessary, those for purely geometrical features of the situation and those for differences in illumination.

Each report of a lookout was corrected according to a number of rules and expressed in terms of a unit named Equivalent Square Feet (abb.: ESF). The unit represented a hypothetical target silhouetted against a sky of brightness equal to .05 foot-lamberts at a distance of 1,000 yards. The unit permitted the comparison of data from many sources and conditions. Despite the many possible sources of error, Wedell and Prentice found that the unit was stable enough to give meaningful results.

The average value for the performance of all lookouts turned out to be very stable at about 60 ESF. Thus the average lookout can see a target of an area of 60 square feet against a bright moonlit sky at 1,000 yards. Against a clear starlit sky at 1,000 yards, the same man requires a target of 1,000 square feet in area. At 2,000 yards in starlight, he requires a target of about 3,800 square feet, or something like a PC boat seen broadside. At 2,500 yards in starlight, he would have trouble seeing a small cargo ship. At the same distance on an overcast night with no moon, he would require a ship the size of a very large freighter. On such nights the maximum useful range of the average lookout is not much over 3,000 yards for even the largest ship targets.

All of these figures are, of course, subject to modification by weather, the lookout's condition, and other factors such as the height of the lookout above the sea. Men higher up have a better chance of seeing the bow wave or wake of fast-moving ships, but.the men lower down will get the greatest advantage from contrast against the sky; the "typical lookout" whose effectiveness is exhausted at about 3,000 yards anyway will be handicapped on dark nights by being stationed high above the water in order to increase the distance of his horizon.

The differences between men are very great. The men differed from one another, when scores on five or more nights were averaged, over a range of 14 ESF to 210 ESF. The poorest man required a target fifteen times the size required by the best man. Thus the poorest man could see a target only when it was at about one quarter of the distance at which the best man could see it. The data suggest that on an overcast night the poorest man can see only the largest passenger liners at 2,000 yards whereas the best man can see ships much smaller than destroyers.

The consistency of the twenty lookouts on whom observations were made on four nights each was indicated by a reliability coefficient of .80.[26] For the ninety-three men who made

---

[26] Odd-even, corrected by the Spearman-Brown formula. The value of the coefficient is that obtained after transformation of the raw data to a logarithmic scale on which the mean score was 1.49 log ESF with standard devia-

reports on more than one occasion the mean range of average scores was 100 ESF. Lookouts who are better (or worse) than average on one occasion may be expected to remain better (or worse) than average on a second occasion, but even a good lookout may be expected to miss fairly easy targets from time to time.

The experience on shipboard suggested the need for more and broader training of night lookouts. L. H. Lanier combined a number of ideas in two synthetic devices for shipboard night vision training.[27] The devices were intended to demonstrate the principles of night vision and to encourage spontaneous practice in use of the eyes at night. One of the two is illustrated in Plates 1 and 2. It was recommended by the Medical Research Laboratory, Submarine Base, New London, for use on all ships of the destroyer escort class and larger. The Training Aids Division of the Bureau of Naval Personnel took no action on this recommendation.

In addition to the night vision trainers the project developed a device to demonstrate the value of night binoculars (always a subject of some controversy in the Navy) and assisted the Navy group at New London in the preparation of the standard manuals for lookout officers and instructors.[28]

tion of .40 log ESF. Without the logarithmic transformation, and before the Spearman-Brown correction, $r=.27$.

[27] L. H. Lanier, *A Night Lookout Trainer for Use Aboard Ship*. OSRD Report 4323. November 8, 1944. Princeton University. Washington, D.C., Applied Psychology Panel, NDRC.

[28] *Lookout: A Manual for Lookout Officers and Supervisors*. NavPers 16198. *Lookout Manual*, NavPers 170069. Washington, D.C., Navy Department, Bureau of Naval Personnel.

CHAPTER 5

THE SELECTION AND TRAINING OF COMMUNICATIONS
AND RADAR PERSONNEL

MODERN war requires the effective coordination of many units through rapid accurate communications. The successful transmission of messages over radio code and voice communication systems is a primary element in winning battles. Closely related to communications in military operations is the kind of radar used in searching for enemy aircraft or submarines and in navigation.

The Applied Psychology Panel studied the selection and training of radio code operators, of radar operators, and of personnel for voice communications. In the last two of these subjects there was a marked tendency to go beyond research on selection and training and to engage in studies of operational procedures.

RADIO CODE OPERATORS

In World War II there was widespread dissatisfaction with the results produced by radio code schools.[1] Hundreds of thousands of men were being trained in radio code each year, and these men were generally not good enough as code operators to handle field communications with the necessary efficiency. Both the Army and the Navy requested Panel assistance in the improvement of the proficiency of radio code operators. Panel research began on October 1, 1942, and continued until the end of the war.[2] G. K. Bennett was in charge of the work and was assisted by the staff shown in Table 19.

In many respects the research program on the radio code

---

[1] This section is based on *STR* I, Chapter 6, and *STR* II, Chapter 12, by Dael Wolfle.

[2] Summaries and complete bibliographies of the research are given in: Albert K. Kurtz and Harold Seashore, *The Radio Code Research Project: Final Re-*

TABLE 19. The research staff of the Panel project on radio code operators.

Contractor: The Psychological Corporation
Contractor's Technical Representative: G. K. Bennett
Project Directors: A. K. Kurtz, F. S. Keller
Staff: F. R. Barrow, G. R. Burns, Reginald Carter, K. W. Estes, M. J. Herbert, E. A. Jerome, H. H. Kendler, C. R. Langmuir, P. G. Murphy, E. J. O'Brien, Cyril Rappaport, H. G. Seashore, S. E. Stuntz, H. R. White, J. M. Willits

---

operator was a model study in military psychology, at least within the limited fields of selection and training. Except for the inevitable delays and confusion produced by wartime interference with the use of soldiers and sailors as subjects of experimentation, the program moved in an orderly systematic manner from general psychological principles to particular applications of proven value. The work began with a thorough survey of radio code training. It continued in the development of a radio code proficiency test which then served as the criterion for the evaluation of psychological tests and training methods. With an acceptable criterion in hand, it was a straightforward technical problem to apply existing knowledge to the improvement of selection and training. A valid independent aptitude test for the radio code operator was prepared. New training methods were developed. A significant by-product of the psychological research appeared in a Morse Code Actuated Printer which was designed as a training device and turned out not only to serve that purpose but to have possibilities as a monitor of radio code circuits. Throughout

port of Project N–107. OSRD Report 4124. September 12, 1944. Psychological Corporation. Washington, D.C., Applied Psychology Panel, NDRC.

Fred S. Keller, *The Radio Code Research Project: Final Report of Project SC–88.* OSRD Report 5379. July 25, 1945. Psychological Corporation. Washington, D.C., Applied Psychology Panel, NDRC.

George K. Bennett, *Research on Project NS–366, Development of Morse Code Actuated Printer: Final Report on Contract OEMsr–830.* OSRD Report 6233. October 31, 1945. Psychological Corporation. Washington, D.C., Applied Psychology Panel, NDRC.

this series of contributions the technical skill of the staff was apparent.

The first problem was to become acquainted with the standard service procedures in selecting and training radio code operators. A survey of twenty schools showed a surprising diversity of organization, methods, and standards. In some of the poorer schools the failure rate was 40 to 50 per cent. In others the poor performance of students was hidden by a confusion in terminology.

The ability of a radio operator was measured by the number of 5-letter words which he could send or receive per minute, but the meaning of "word" varied from school to school. The word as a unit of measurement was applied to plain English text of varying degrees of familiarity, to code words, to numbers, and to mixed letters and numbers. One, five, ten or more minutes of actual practice might be used in measuring performance, and a man might be said to be able to receive code at so many words per minute only when he received it perfectly or he might be allowed 5, 10, 20, or even 40 per cent errors. The combined effect of these variations was enormous and hid the effects of good and poor methods of selection and training.

The Panel took two special steps to reduce the diversity of radio code school practices. The first was to prepare a monograph on the measurement of radio code speed.[3] This brought together the numerous concepts and gave detailed instructions for the conversion of measurements made under one set of conditions to measurements made under other sets of conditions. The second step was to develop a standardized proficiency test. One test of operator performance in receiving code was developed for research purposes. A second test was prepared later for use in training men.

The most difficult code school problems seemed to center

[3] Albert K. Kurtz, Harold Seashore, Stephen E. Stuntz, and John M. Willits, *The Standardization of Code Speeds*. OSRD Report 3490. May 1944. Psychological Corporation. Washington, D.C., Applied Psychology Panel, NDRC.

on the ability to receive code; therefore the Code Receiving Tests[4] were developed to measure the speed and accuracy with which a trained radio operator could copy the alphabet and numbers of the International Morse Code. All material for these tests was recorded on phonograph records. The test items consisted of five-letter mixed-code groups in which each of the twenty-six letters and ten numbers appeared with equal frequency. A test was constructed for each of a number of code speeds ranging from very slow to rather fast. Thus progress in learning to receive code could be measured. There were two equivalent forms of the test for each speed. The reliability (agreement between alternative forms) of the test at various speeds was .93 to .96 when administered to servicemen at appropriate stages of training.

The Code Receiving Tests were widely used as a research tool to measure the progress of men being trained in different schools, under different conditions, and at different times. For this purpose the tests were excellent, but they were not designed for use as regular school examinations. For this purpose the Radio Code Receiving Examinations were constructed by the project on Navy Aptitude Tests in connection with its general program of achievement and proficiency testing (see Chapter 8). The Radio Code Receiving Examinations, complete with practice units and directions to the testees, were recorded on phonograph records and issued by the Bureau of Naval Personnel for all its code schools.

For the selection of radio code operators special aptitude tests were desired. Three chief types of radio code aptitude tests had already been tried: (1) Discrimination tests, which measure the ability of the subject to distinguish between complex rhythmic patterns of dots and dashes. The best-known example of this type is the Signal Corps Code Aptitude Test,

[4] Albert K. Kurtz, Harold G. Seashore, and John M. Willits, *The Experimental Edition of Code Receiving Tests*. OSRD Report 1314. March 29, 1943. Albert K. Kurtz and Harold Seashore, *The Code Receiving Tests*. OSRD Report 3157. February 2, 1944. Psychological Corporation. Washington, D.C., Applied Psychology Panel, NDRC.

which was in common use at the beginning of the war. (2) Code learning tests, which measure the ability of the subject to learn code characters. A test developed by L. L. Thurstone and an Army test (ROA-2) are examples. (3) Speed of response tests, which measure the ability to receive a few easily learned code characters at rapid rates of transmission. The NDRC Speed of Response Test[5] developed for the Panel by A. K. Kurtz is the best modern example.[6]

The Speed of Response Test has two parts. First there is a learning period in which a few code characters are taught. Second there is a testing period in which the subjects are tested on these learned characters at increasingly faster rates of presentation. The score represents the total number of characters correctly received during the testing period. Separate forms of the test were prepared for Army and Navy use. After validation these became the official code aptitude tests of the Army and Navy and were adopted also by a variety of associated war activities.

The validity and other characteristics of the Speed of Response Test were determined in a series of Panel studies conducted on service personnel. Independently, L. L. Thurstone of the University of Chicago, the Army, and the Navy compared the test with other tests for the prediction of performance in learning to receive code. The validity coefficients obtained are listed in Table 20.[7] In the whole series of studies the Speed of Response Test showed validity correlations ranging from .30 to .61. In direct comparison it was equal in validity to the Thurstone test and required less time to administer. It was superior in validity to the other code tests and far superior to the tests of the U.S. Navy Basic Classification Test Battery. Its interrelations with the tests of the Basic

---

[5] Albert K. Kurtz, *The Prediction of Code Learning Ability.* OSRD Report 4059. August 26, 1944. Psychological Corporation. Washington, D.C., Applied Psychology Panel, NDRC.

[6] The idea for this type of test came from R. A. Biegel, Eine Eignungsprüfung fur Funkentelegraphisten. *Psychotechn. Z.* 1931, *6,* 41–45.

[7] The data of Table 20 are summarized from Tables 1, 2, and 3 of *STR* I, Chapter 6, by Dael Wolfle.

Battery were very low, as were its interrelations with age, education, and clerical ability.

TABLE 20. The validity of the Speed of Response Test (abb: SOR) and three other tests of radio code aptitude.

Test	Study by	Criterion	Number of Classes	N	Weeks of Instruction	Median Validity Coefficient
SOR	Thurstone	Code speed	1	215	6	.61
SOR	Thurstone	Pass–fail in course	1	196	16	.57
SOR	Army	School speed	4	221	7–8	.55
SOR	Army	Code Receiving Test	4	221	7–8	.41
SOR	Navy	Code speed	1	*	14	.37
SOR	Navy	Pass–fail in course	1	*	14	.49
SOR	NDRC	Code Receiving Test	2	216	5	.30
SOR	NDRC	Code Receiving Test	10	1932	10	.34
Thurstone	Navy	Code speed	1	*	14	.36
Thurstone	Navy	Pass–fail in course	1	*	14	.42
Thurstone	Army	School speed	4	221	7–8	.47
Thurstone	Army	Code Receiving Test	4	221	7–8	.44
ROA X–1	Army	School speed	4	221	7–8	.33
ROA X–1	Army	Code Receiving Test	4	221	7–8	.39
SCCAT, I	Army	School speed	4	221	7–8	.30
SCCAT, II	Army	School speed	4	221	7–8	.33

* For the Navy studies N varied from 183 to 306.

Since verbal and mechanical aptitude, youth, intelligence, education, and clerical ability are all at a great premium for many military assignments, it is highly advantageous to have a test for code aptitude which is valid yet independent of all these. The Speed of Response Test was the best example of a special test produced by the Panel during the war. Its success points to the probability that the problem of test independence, discussed in Chapter 3, can be solved at least for some situations.

The Speed of Response Test was later shown to have almost as high validity for blinker (visual) code as for radio code.[8]

[8] Irving H. Anderson *et al.*, *Experiments in Training Radar Operators in Visual Code Reception.* OSRD Report 4811. March 20, 1945. Yerkes Laboratories. Washington, D.C., Applied Psychology Panel, NDRC.

This suggests that the effect of adapting the presentation to a visual one should be studied.

Radio code training consists of two phases. First, the characters representing the thirty-six letters and numbers must be learned. In the second, speed in recognition of the characters must be acquired. Simultaneously, skill in sending them must be developed. This phase may continue until any desired speed is attained or until the operator reaches his speed limit. In each phase the Panel project made a number of experimental contributions which turned out to have high practical value. A sample of three studies will be described.

The Code-Voice Method of Teaching Radio Code. The first task in learning code is to learn to identify each of the thirty-six sound patterns (dots and dashes) which represent the thirty-six letters and digits. Various methods have been used to produce this learning. Essentially, all consist of sounding the characters and helping the student to associate the sound with its more familiar English equivalent. Thus when the pattern "dot-dash" is sounded, the student may be told that it represents Able (the commonly used word which identifies the code letter *A*), or he may search through a printed list until he finds the correct letter, or he may have to wait until he has heard a number of such patterns before he is informed of the equivalent of each.

On the basis of the general principles of learning, F. S. Keller predicted that learning would proceed most rapidly if the learner was told immediately after hearing each character just what that character was. In order to provide the learner with an opportunity to recognize the characters for himself, a pause of about three seconds was introduced between sending the character and naming it. In order to allow a fairly large number of correct recognitions, the characters were frequently sent in "doubles," for example, dot-dash, pause, "Able," pause, dot-dash, pause, "Able." Then another character was sent, named, repeated, etc. This is known as the code-voice method of teaching code. It is closely related to the paired associates procedure of memorizing verbal material and, like the paired associates procedure, provides knowledge-of-results training.

Together with the psychological advantages of immediate knowledge of results, Keller combined in the code-voice method the principles of whole learning, meaningful scores, a running record of improvement, and the opportunity to compete with one's own past record. With the help of Don Lewis, Panel liaison officer from the Signal Corps, Keller and the project staff were able to test the validity of the code-voice method.[9]

Inexperienced students of three beginning code classes at Camp Crowder, Missouri, were trained by standard Army procedures. Inexperienced students of the next three entering classes at Camp Crowder were given code-voice training for six days. Both groups were tested at the level of five words per minute on the seventh day. Progress in code receiving was measured in terms of the percentage of men who qualified at five words per minute by the end of the seventh day of training. There were 87 men in the standard group and 262 in the code-voice group. The two groups were equal in general ability as measured by the Army General Classification Test and in code ability as measured by the Army Radio Operator Aptitude Test.

The result of the experiment was proof of the superiority of the code-voice method over the standard method. The percentage of standard students who qualified at five words per minute by noon of the seventh day of instruction was 28.74; for code-voice students, the value was 50.00. Statistically the difference was highly significant. Other comparisons of the two groups gave further confirmation of the value of the code-voice method.

[9] Fred S. Keller, Katherine W. Estes, and Paul G. Murphy, *A Comparison of Training Methods at Two Levels of Code Learning*. OSRD Report 4329. November 10, 1944. Fred S. Keller, *Memorandum on the Code-Voice Method of Teaching International Morse Code*. OSRD Report 4911. April 9, 1945. Psychological Corporation. Washington, D.C., Applied Psychology Panel, NDRC.

The experiment led to the adoption of the code-voice method at Camp Crowder. The Panel cooperated with the Radio Training Section of Camp Crowder in preparing an instructor's manual for the method and an introductory pamphlet and explanation for the student. The method and instructions for teaching code by the code-voice method were later incorporated into the general War Department manual[10] which describes the doctrine for training the radio code operator.

The Distribution of Practice in Training Code Operators. Following the learning of the individual characters making up the International Morse Code, the student must learn to identify each character as it is sent to him at faster and faster rates. Acquiring this skill takes up most of the time spent in code school. It is a relatively slow and sometimes tedious process.

In an effort to force students to attain higher operating speed, men in the code school at Camp Crowder were given seven hours a day of code instruction. To Keller it seemed likely that this schedule was accomplishing no more than could be accomplished in a shorter work day. In a variety of previous psychological studies it has been shown that increasing the daily hours of drill or work beyond some optimum leads to decreased learning or decreased output. An experimental investigation was therefore made of the relative value of seven hours per day of practice in code reception as against four hours per day.

In the normal schedule at Camp Crowder, students practiced seven hours a day, five days a week, and four hours a day on Saturday for the first five of the eight weeks of code school. The final three weeks were devoted to other topics. Code learning in a group of 355 men working on this schedule was compared with that of 165 men who had their training spread over the entire eight weeks on a schedule of four hours per

[10] *International Morse Code (Instructions).* TM 11–459. August, 1945. Washington, D.C., War Department.

day, six days per week. The groups were equated for code aptitude and Army General Classification Test scores.

The results of the comparison are shown in Figures 8 and 9. Each bar of Figure 8 indicates the per cent of students in a group who had mastered a given code speed at the end of the first five weeks. Despite the reduction of three hours per day of practice the group on a four-hour schedule did just as well as the group on a seven-hour schedule.

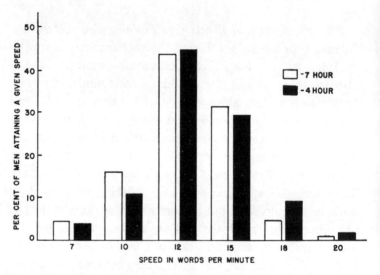

Figure 8. Ability to receive radio code after five weeks of training on a schedule of seven hours per day in contrast to four hours per day. Each bar shows the proportion of a group who could pass a test of code reception at a given speed. Since the four-hour group did as well as the seven-hour group, three hours per day were wasted in the training of the latter.

The schedule of the four-hour group was maintained from the fifth to the eighth week of the course. During this time the seven-hour group practiced code procedure, studied equipment, etc. At the end of eight weeks the four-hour men were retested. As shown in Figure 9 the four-hour group had attained an increased mastery of reception at higher speeds. Thus the experiment showed that either three hours of each day in the first five weeks of code-school training could be devoted to other aspects of code-training than the reception of code, or increased mastery of radio code could be attained by spreading the total number of training hours over the whole eight-week period.[11] Supplementary comparisons within the four-hour group showed that it made no difference whether the four hours of training were massed in one morning session or distributed over hours 1, 2, 4 and 7 of the day.[12]

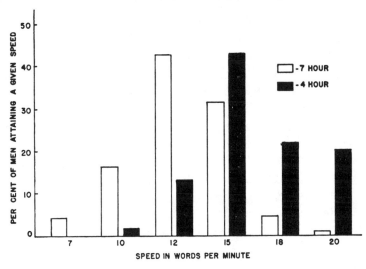

Figure 9. Ability to receive radio code after eight weeks of training on the two schedules of seven and four hours per day. See text for explanation.

The Morse-Code Actuated Printer. One of the difficulties of radio communication is poor sending. Rating the quality of sending is not easy. A technique frequently employed in code schools is to record the copy sent by the student. The student is then required to "read" his own sending as it is played back to him. In other schools, the instructors or fellow

[11] Fred S. Keller and Katherine W. Estes, *The Relative Effectiveness of Four and Seven Hours of Daily Code Practice.* OSRD Report 4750. February 26, 1945. Psychological Corporation. Washington, D.C., Applied Psychology Panel, NDRC.

[12] Fred S. Keller and Katherine W. Estes, *Distribution of Practice in Code Learning.* OSRD Report 4330. November 10, 1944. Psychological Corporation. Washington, D.C., Applied Psychology Panel, NDRC.

students read the students' sending. All of these techniques involve variable standards of excellence. An *A* (dot-dash) may be understood as such even though the dash is drawn out to greater than normal length. A *5* may be understood as such even though it contains six or seven or even eight dots instead of the correct five. In plain English copy the context helps the receiver greatly so that perfect sending is not always necessary. The fact that much less than perfect sending can sometimes be tolerated results in sending which is generally poor. Erratic sending can be expected to produce many errors when difficult conditions are met in the theater of operations.

Applied Psychology Panel personnel working on problems of code training agreed that a trainer which would give an objective record of the quality of a student's sending would be an aid to instruction. Such a trainer was built by G. K. Bennett and C. R. Langmuir.[13] It consists of an electric typewriter which is controlled by discriminator and analyzer circuits in such a way that radio code signals fed into the device are printed as ordinary typed characters on the typewriter.

The Morse-Code Actuated Printer was designed primarily as a trainer. If a man sends a letter correctly, that letter is printed by the typewriter. If his sending is incorrect, the nature of the typed character, or characters, indicates the nature of the error made. If, for example, in attempting to send the character *H* (four dots), the student sends five dots, the typewriter prints *5*. If the fourth dot is prolonged into a dash, the typewriter prints *V*. If the pause between the second and third dots is made too long, the typewriter prints *II*. In each case, if an error is made, the nature of the printed record gives both student and instructor an immediate indication of the meaning of that error. If the sending is perfect, within adjustable tolerance limits, the student has the satisfaction of seeing perfect copy appear on the typewriter. Operators with

[13] George K. Bennett, *Research on Project NS–366, Development of a Morse Code Actuated Printer: Final Report of Contract OEMsr* 830. OSRD Report 6233. October 31, 1945. Psychological Corporation. Washington, D.C., Applied Psychology Panel, NDRC.

long experience have tried out the trainer. Their usual comment after a few trials is that it cleans up their sending better than anything they have ever seen.

Since the Morse-Code Actuated Printer takes the code off the air as well as from training circuits, it may also be used to monitor circuits or as part of a communication net. In such uses it has an operating speed of up to 100 words per minute. Plate 3 reproduces a sample of a tape record taken off the air together with a picture of one model of the Morse-Code Actuated Printer itself.

Three different models of the printer were constructed and turned over to the Navy Bureau of Ships for service trial. The commercial uses of the Morse-Code Actuated Printer are under exploration now that the Panel work has terminated.

The research of the Panel on radio code illustrates the values of military psychology at their highest in so far as problems of classification and training are concerned. There need be only one qualification to this generalization; much of the research was completed too late to have a maximum effect in World War II. Since nearly all of the research could have been done years ago, and since all of it was accomplished in the face of the constant interruptions forced by the war emergency, the qualification suggests the need for continuous research on this and similar problems with more favorable timing and better research conditions. It is conservative to say that intelligent systematic administration, from the beginning of the war to the end, of the principles and materials developed by the Panel for radio code alone would have saved our country thousands of man-years while sensibly increasing our combat efficiency.

Radar Operators

One of the major developments of World War II was Radio Detection and Ranging—radar. By the winter of 1943 the applications of radar had become numerous and diverse. The task confronting an operator varied from simple detection of targets and ranging on them to multiple and simultaneous functions requiring discrimination and interpretation of target signals in addition to the determination of target bearings and altitudes. The conditions under which these functions were performed ranged from the comparative security of the ground-based coastal search station to the bombing mission over enemy territory.[14]

In operating a radar set attention is centered on the visual presentation of the radar oscilloscope tube. Some radar sets displayed a moving bobbing pip which had to be adjusted to a hairline. Other sets showed two pips which had to be adjusted to equal one another in height. In still other sets there was a line which moved horizontally or radially, leaving behind it momentary bright spots which represented objects in the field of view of the radar antenna.

Whatever the primary presentation, there were likely to be secondary disturbances on the tube face, false echoes, for instance, or random brightening of the tube face due to amplifier noise. Interference lines and patterns from neighboring radar sets or from deliberate enemy "jamming" also occasionally appeared on the tube. Stray light on the surface of the operator's own tube sometimes gave glare spots which increased the difficulty of seeing the target spot.

In addition to attending to the visual presentation of target signals, the radar operator read dials or grid coordinates, adjusted a variety of tuning, calibrating, and operating controls, and cooperated with his team-mates. Cooperation was especially complex in the case of radar bombing where radar operator, bombardier, and pilot all had to coordinate their work. As the war progressed there was a tendency to reduce the need for cooperation in elementary operations, partly by making some functions automatic and partly by requiring one operator to perform more functions.

In the early days of the war there was little attention in this country to the problems of the radar operator. Except in

[14] This section is based on *STR* I, Chapter 7 and *STR* II, Chapter 2, by D. B. Lindsley.

the case of the adaptation of radar to the needs of the aircraft pilot in night operations, interest centered on the development of instruments of longer range and greater precision and in the training of thousands of men to maintain these new and complex electronic devices. It was thought that greater gains were to be achieved by new materiel discoveries than by securing optimum use of materiel which was hardly produced before it became obsolete. Neverthless, thousands of men were trained to operate and did operate in combat the very equipment which already was regarded as obsolete by laboratory technicians. Hence the NDRC Radiation Laboratory at Massachusetts Institute of Technology and other NDRC sections began to build synthetic radar training devices. In addition the Applied Psychology Panel undertook to study the Selection and Training of Oscilloscope Operators.[15] The work of the Panel began in February 1943 upon the request of the Signal Corps and the Navy Office of the Commander in Chief, U.S. Fleet. Don Lewis and Capt. D. C. Beard, USN, respectively Army and Navy liaison officers, made notable contributions to the work. D. B. Lindsley, for the Panel, was assisted by the staff shown in Table 21.

TABLE 21. The research staff of the Panel project on radar operators.

Contractor: Yerkes Laboratories of Primate Biology
Contractor's Technical Representative: H. W. Nissen
Project Director: D. B. Lindsley
Staff: I. H. Anderson, A. L. Baldwin, C. S. Bridgman, R. S. Daniel, J. G. Darley, R. E. Dreher, E. P. Horne, E. A. Jerome, W. H. Lichte, T. L. McCulloch, Fred McKinney, K. U. Smith, G. R. Stone, E. J. Sweeney, G. J. Thomas

[15] Because the work was anticipated to be in Florida, the Panel asked the Yerkes Laboratories of Primate Biology to serve as a contractor of convenience. H. W. Nissen unselfishly contributed a great deal of his time as contractor's technical representative. A summary and complete bibliography of the work of the project is given in D. B. Lindsley, *Final Report in Summary of Work on the Selection and Training of Radar Operators.* OSRD Report 5766. September 24, 1945. Yerkes Laboratories. Washington, D.C., Applied Psychology Panel, NDRC.

The project's analysis of the tasks of the radar operator suggested that a critical aspect was visual perception, especially the ability to detect visual changes quickly and accurately and to note movement, form, and spatial relationships. A series of paper-and-pencil tests for these capacities was constructed. The reliabilities of the tests were reasonable, and the tests had high intercorrelations with one another. With AGCT the intercorrelations were between .40 and .50 in all but a few cases. Nearly all of the tests simulated one or more phases of the operator's task, so that they had high face validity. Securing proof of true validity was difficult because of the absence of reliable criteria. After the Panel's preliminary studies[16] had been completed, the Bureau of Naval Personnel revised several of the tests for inclusion in its definitive radar battery. Following the revision, validities of .40 to .50 were obtained for the battery in the prediction of radar school grades. The Air Surgeon also revised several of the tests for use in the selection of radar bombardiers.

Studies of visual tests for radar operators[17] suggested that poor visual acuity in both eyes at near distances and overconvergence at near distances as measured on the Bausch and Lomb Ortho-Rater may be associated with poor proficiency in oscilloscope operation. Operators who were superior in these visual functions were rated by their officers as slightly better than operators who were inferior.

The sudden and large demand for radar operators, both in the Army and in the Navy, meant the rapid establishment of radar operator schools, the selection of instructors from relatively inexperienced personnel, the securing and setting up of new equipment for demonstration and training purposes, and the graduation of the large numbers of operators required to

[16] Irving H. Anderson et al., A Validational Study of Oscilloscope Operator Tests. OSRD Report 3712. April 24, 1944. Yerkes Laboratories. Washington, D.C., Applied Psychology Panel, NDRC.
[17] Irving H. Anderson et al., Vision as Related to Proficiency in Oscilloscope Operation. OSRD Report 3409. February 24, 1944. Yerkes Laboratories. Washington, D.C., Applied Psychology Panel, NDRC.

meet quotas. The research staff visited a number of Army and Navy schools and made suggestions for improving the training programs. Since it was typical of many, the results of one survey at the Navy school at Virginia Beach, Virginia, will be summarized. I. H. Anderson, J. G. Darley, and W. H. Lichte were generally responsible for this type of work of the project.

The length of the course at Virginia Beach was three weeks. During this time the trainee was expected to become familiar with the operation of several types of radar gear. In addition, the trainee was required to know something about target detection and interpretation, interference, jamming and antijamming, aircraft control, navigation, dead-reckoning plotting, and intercommunication procedures. Obviously only a smattering of knowledge on any of these topics could be acquired in the time available for training. Included in the project's comments on the Virginia Beach program were the following:

1. The lectures and the operator's handbook should place more emphasis on practical and functional aspects of operation and on operator skills. Many instructors had the notion that an operator had to know radar theory and maintenance. Knowledge of theory and maintenance would have been valuable but it could not have been given in three weeks even if the whole period had been available for training as a radar mechanic.

2. Laboratory and operational procedures should have adequate supervision. Students were assigned to operate equipment with insufficient knowledge of correct operating procedures. Thus valuable time assigned to drill was wasted by trial-and-error procedures which often resulted in poor habits of operation.

3. Laboratory work should take up one kind of equipment at a time. Operators were shifted from one gear to another before completely mastering any one. This led to confusion and failure to assimilate any specific operational procedure.

4. Plotting and operating periods should be organized with

specific problems in mind. They should not rely on rote memory of procedures and on random operation of radar gear. Drill problems should stress tactical and functional uses of the gear.

5. Greater emphasis should be placed upon speed and accuracy in operation. The correct procedure should be taught at the start of training. When an operator has mastered the correct procedure, gradual increase in speed of operation should become the goal.

6. An operator should be taught that calibration and tuning are basic to efficient operation and that the greatest care must be observed in carrying out these procedures.

7. Greater recognition should be given to the fact that radar operation is a *skill* which may be developed to a high degree. Knowledge of operating procedure is not enough. Skill comes only with practice in the job. Therefore every opportunity should be given the trainee to exercise his knowledge by operating the gear. The timing of performance and posting of operational scores is a good way to foster competition and provide motivation to work harder at the task of improving skill and speed.

8. With better information concerning eventual assignment of trainees to duty, it should be possible to determine the type of equipment needed for the proper training of each man. This would reduce the number of different types of gear upon which an individual operator might be given training. By concentrating on a few types, much greater proficiency could be attained.

9. Greater use should be made of training aids, especially visual aids. Models, wall diagrams, illustrations, slides, and movies are indispensable for demonstrating the correct procedures or the basic principles behind operating procedures. These should be coordinated with assignments, lectures, and demonstrations.

10. Greater use should be made of radar trainers and similar devices by which operating problems may be presented,

specific components of operating procedures may be illustrated and practiced, and quantitative scores on performance may be obtained.

11. More time and attention should be given to target interpretation and the dynamic aspects of oscilloscope viewing and reading. An effort should be made to illustrate all conceivable types of targets, land masses, fading, and other phenomena which may occur during operation. Operating procedures for all possible exigencies should be practiced. The basic principles of oscilloscope interpretation, distortion, and interference should be made concrete.

12. More attention should be given to the development of objective measures of proficiency. Such measures should emphasize the performance aspects of operation and the solving of realistic operating problems.

13. Fleet operational requirements constantly change to meet the demands of new tactics. In order to keep the training in the schools up to date instructors should be selected for a period of sea duty, rotating with men who have had shipborne operational experience. This system would bring the latest operational procedures into the schools and bolster the morale of the trainees.

Many of the detailed recommendations which supported these generalizations were adopted. Similar studies occurred in other Army and Navy schools with similar results. The surveys of training in low-altitude radar bombing, ground-controlled-approach systems, submarine radar,[18] and other fields resulted in the preparation by the project of achievement and proficiency tests, workbooks, training manuals, synthetic trainers, and training aids.

Achievement and Proficiency Tests. One of the conclusions of the training surveys was that there was a definite need for the improvement of examination methods. In particular it was deemed advisable to develop objective examinations as measures for use at the end of the training period (see Chap-

[18] In cooperation with Divisions 6 and 17, NDRC.

ter 8 for a general discussion of such examinations). Many achievement and proficiency tests for radar operators and radar mechanics were developed by I. H. Anderson, A. L. Baldwin, and J. G. Darley.

This type of work required extensive work with each school, since it was necessary to prepare objective examinations for a number of different types of radar gear. In a series of studies for the Bureau of Naval Personnel, for example, nineteen experimental tests were made and tried out in various schools. An item analysis was made to rule out items which did not contribute significantly to the total examination. Finally, three radar operator final achievement batteries of two forms each were prepared and submitted to the Bureau of Naval Personnel for reproduction and use in the schools. Each examination included sections on basic radar, two or more types of radar gear, dead-reckoning-tracer plotting, relative motion, navigation, air plotting, and surface plotting.

Synthetic Trainers. There were many tactical operating problems of radar which were difficult to teach in training. Often these could be made vivid and real by means of trainers which produced synthetic targets and which allowed not only maneuvering of targets but also made provisions for scoring the performance of the operator. The project developed several such devices, including a PPI Flash Reading Trainer,[19] a Mechanical PPI Tracking Trainer,[20] and several motion-picture trainers for radar bombing. R. S. Daniel and K. U. Smith were most active in developing the Panel's radar trainers.

The PPI, or Plan Position Indicator, displayed a kind of map of the territory around the radar set. Objects appeared on this map as bright spots on a dark background. The spots

[19] Irving H. Anderson *et al., A Radar Trainer and Flash-Reading Method for Operators of the Plan Position Indicator.* OSRD Report 4831. March 20, 1945. Yerkes Laboratories. Washington, D.C., Applied Psychology Panel, NDRC.
[20] D. B. Lindsley, *A New Type Mechanical PPI Tracking Trainer.* Project SC–70, NS–146, Informal Memorandum No. 26. September 4, 1945. Yerkes Laboratories. Washington, D.C., Applied Psychology Panel, NDRC.

varied in brightness and duration. One task was to read target position in terms of grid coordinates on the face of the tube. The project's aptitude tests suggested that the average American selected for training as a radar operator would be rather inaccurate in reading grid coordinates. The need for a trainer to give well-motivated drill was indicated, and R. S. Daniel built the project's PPI Flash-Reading Radar Trainer. With it target spots could be presented on five student mock oscilloscopes simultaneously. The order, timing, position, and persistence of the targets could be varied at will and provision was made for a simple scoring system.

As typified by the PPI Flash-Reading Trainer, the Panel's approach to radar trainers was conceived as a supplement and simplification of the more general NDRC program on radar trainers. Other NDRC groups built synthetic devices with electronic controls for most of the later radar sets as the new sets were devised. Although electronic trainers gave very realistic simulations of operator tasks, their production interfered with the production of equipment for radio and radar sets in general. The Panel attempted to avoid this interference by designing optical and mechanical rather than electronic systems for its synthetic radar devices. Optical and mechanical systems were sometimes less realistic but, by a judicious selection, the subordinate unit tasks were presented in ways believed to be sufficiently realistic.

The project also found that a surprising degree of dynamic simulation could be provided by means of problems laid out on successive pencil-and-paper diagrams. A great variety of such problems was developed in the tests, workbooks, and training manuals prepared by the project for use in the Army, Navy, and Air Forces.

Another way in which practice in the interpretation of targets could be made possible was through the use of a series of photographs, or better, a movie of actual oscilloscope presentations. Such pictures were taken in the air by Lindsley and Daniel. The still pictures were arranged in a problem series requiring the trainee to work out a solution. The movies were presented on a mock radar oscilloscope. Although they were originally intended for training, it was obvious that the pictures could readily be extended to the more immediate preparation for combat in briefing operators before bombing raids. As the war ended steps were being taken to secure general use of motion pictures of actual targets in briefing.[21]

Throughout the life of the project every effort was made to complete validational studies of synthetic trainers. In general this proved to be as hard as the validation of radar aptitude tests. In consequence, both for its own and other trainers, the project's studies were limited to empirical trial of the devices in use and to the determination of learning curves.[22] The latter permitted a rational choice of the number of hours to be devoted to the trainer in the schools.

An Experiment on Training in Bombing. When radar bombing began in the active theaters there were numerous complaints about its accuracy. It was not known whether this was due to the type of equipment being used or to insufficient length of operational training. In order to solve this problem the Army Air Forces Training Command set up an experiment on the effects of extended training. The Panel was asked to supervise and report upon the results of this experiment. A. L. Baldwin and D. B. Lindsley represented the Panel.

The study was carried out during the period from April to July 1945. During this time twenty graduates of the regular course in radar bombardment, including ten bombardier-trained operators and ten navigator-trained operators, were given roughly five times as much operational practice as was customary. Bombing proficiency was scored by a photo-

[21] D. B. Lindsley, *Radar Scope Movies for Briefing and Reconnaissance Purposes.* Project SC-70, NS-146, Informal Memorandum No. 30. September 24, 1945. Yerkes Laboratories. Washington, D.C., Applied Psychology Panel, NDRC.

[22] See, for instance, Irving H. Anderson *et al.*, *A Study of the SCR-584 Basic Trainer as a Training Device for Learning Range Tracking.* OSRD Report 3344. February 10, 1944. Yerkes Laboratories. Washington, D.C., Applied Psychology Panel, NDRC.

bomb-scoring method and a detailed analysis of the sources of error was made.

The experiment showed that between three and four times the usual operational training period was required to reach peak efficiency when the total group of operators was considered. The ten navigator-trained operators required only double the usual length of training to reach peak efficiency, thus raising a question concerning the adequacy of selection of the men to be trained as radar bombardiers.

An analysis of the sources of bombing errors showed that oscilloscope interpretation was most responsible and that more emphasis should be placed upon training in target analysis and oscilloscope interpretation. The next most important source of error was faulty set calibration.

During the course of the extended training there was a consistent reduction in the circular error, range error, and deflection error of bombing. The results of the experiment provided information on the general level of bombing accuracy which could be expected with the particular kind of equipment used, and clearly indicated that with additional training of the operational type much greater proficiency could be attained.

In addition to its research on selection and training, the project completed several series of studies on operating procedures for radar. The topics investigated included the effects of ambient light on pip detection, visual code messages, the distribution of working periods, the methods of tracking targets in range, and the effects of operation of radar sets on human vision. In the case of the last it was shown that many months of radar operation had no deleterious effects on ability to pass the visual tests of the Ortho-Rater.[23]

The work of Lindsley and his colleagues was of consider-

[23] Irving H. Anderson *et al., Effect of Oscilloscope Operation on Vision.* OSRD Report 2051. November 15, 1943. *Visual Status of ASV Radar Operators.* OSRD Report 3443B. March 20, 1944. Yerkes Laboratories. Washington, D.C., Applied Psychology Panel, NDRC.

See also J. K. Adams, D. C. Beier, and H. A. Imus, *The Influence of the Visual Tasks Required of Personnel in the 16 Weeks Fire Controlmen (O) Training Course upon Their Visual Proficiency.* OSRD Report 3970. August

able service to the Army and Navy, particularly in the matter of the training of radar operators. Their work illustrates at its best the results which may be produced by independent consultants who are thoroughly familiar with service problems and able to spend whatever amount of time is required to assist in detailed application of their recommendations.

VOICE COMMUNICATION

Military operations require teamwork, and teamwork requires rapid accurate communications. For rapid and complete understanding the most satisfactory kind of communication is the voice—the telephone or radio voice when direct communications are impossible.[24]

The use of the telephone is so common that it seldom occurs to anyone who is unfamiliar with military conditions that there is any problem other than distance to be overcome in military communications by voice. Yet military operations are always marked by noise. The noise of battle is only the extreme case. Military equipment is noisy, men at work are noisy, the wind and rain and waves create noise.

When a microphone is used in noise, the noise enters the circuit and disturbs the listener. When earphones are worn in noise, the listener always finds that the noise gets into the ear and conflicts with the voice over the wires. The radio is likely to be worse than the telephone because static, the simultaneous operation of friendly units, and deliberate interference by the enemy add to the noise.

Faulty communications endanger military operations. Yet in a considerable number of cases voice messages must be accurately sent and correctly interpreted under conditions roughly analogous to the use of a telephone in a fast-moving subway train with the windows opened.

Many partial solutions of the problem are possible. Among

1, 1944. University of Wisconsin. Washington, D.C., Applied Psychology Panel, NDRC.

[24] This section is based on *STR* I, Chapter 10, and *STR* II, Chapters 10 and 11, by Dael Wolfle.

them two were undertaken by NDRC. Stated in its most general terms, the first was to increase the fidelity of the system and the ratio of signal to noise. The second was to increase the ability of personnel to understand communications.

Even slight improvements in fidelity or in the signal-to-noise ratio produce considerable gains in understanding. Modification of communication equipment was undertaken by several NDRC divisions. In many cases the results were tested for their effects on understanding by psychologists under S. S. Stevens of the Psycho-Acoustic Laboratory, Harvard University, a contractor of Division 17, NDRC. The result of this early combination of engineers and psychologists was a steady improvement in communications equipment. Even in the early experiments, however, Stevens pointed repeatedly to the fact that speakers over a given communication system differed in their ability to get calls through, listeners differed in their ability to understand a given speaker, and both speakers and listeners improved with experience in communicating through noise.

In the fall of 1942 Comdr. P. E. McDowell, USN, informally asked the Panel to follow up on Stevens' observations by formulating an action program for the selection and training of "battle telephone talkers," as the personnel who handled telephones on shipboard were called. At about the same time Don Lewis of the Operational Research Branch of the Office of the Chief Signal Officer approached the Panel with an informal request for research on the training of aircrew in voice communications.

The Panel surveyed the possibilities for research,[25] and the opportunities were found to be striking. In the Navy battle telephone talkers were assigned to this duty chiefly because their regular station was convenient to a telephone. As a result time was lost in ships' training. Men who had never before used a telephone, men with marked foreign or regional dia-

25 The survey was made by Grant Fairbanks under the contract with the National Academy of Sciences.

lects, and men with definite speech defects were assigned to battle telephones. By the time the effects of such men on coordination in operations were discovered, the men had already learned their primary duties. Transfer to another duty created its own inefficiencies.

In the Army Air Forces the situation was different in so far as the possibilities for selection of men were concerned. Aircrew personnel were highly selected on education and on various test scores which seemed to insure familiarity with telephones, relative speed in adapting to noisy conditions, and a minimum of speech peculiarities.

In both services, however, it was generally assumed that all men would learn without formal training to adapt to the conditions affecting communications. A program of selection and training was undertaken in the Navy, and a training program combined with research on operating procedures was undertaken in the Army.[26] G. K. Bennett was placed in general charge of the two programs. J. C. Snidecor and, later, L. A. Mallory were responsible for detailed direction of the work in the Navy. J. W. Black was similarly responsible for the Panel's research in the Army Air Forces. The complete staffs of the two projects are shown in Table 22.

Telephone Talkers

A method of selecting telephone talkers[27] was developed in connection with the whole program of classification of the personnel of the U.S.S. *New Jersey* (see Chapter 3). A "speech

[26] Summaries and complete bibliographies of the two projects are to be found in: Louis A. Mallory and William J. Temple, *Final Report in Summary of Work on the Selection and Training of Telephone Talkers.* OSRD Report 5497. August 27, 1945. Psychological Corporation. Washington, D.C., Applied Psychology Panel, NDRC. And also John W. Black, *Final Report in Summary of Work on Voice Communication.* OSRD Report 5568. September 11, 1945. Psychological Corporation. Washington, D.C., Applied Psychology Panel, NDRC.

[27] Anonymous, *A Speech Interview for the Selection of Telephone Talkers.* OSRD Report 1769. August 1943. Psychological Corporation. Washington, D.C., Applied Psychology Panel, NDRC.

TABLE 22. The research staffs of the Panel projects on telephone talkers and aircrew using voice communication systems.

Contractor: The Psychological Corporation
Contractor's Technical Representative: G. K. Bennett

TELEPHONE TALKERS

    Project Directors: J. W. Black, J. C. Snidecor, and L. A. Mallory

    Staff: T. G. Andrews, J. F. Curtis, Grant Fairbanks, H. K. Fink, E. L. Hearsey, G. W. Hibbitt, H. M. Mason, D. G. Powers, W. J. Temple, A. G. Wesman

AIRCREW USING VOICE COMMUNICATION SYSTEMS

    Project Director: J. W. Black

    Staff: I. P. Brackett, J. F. Curtis, G. L. Draegert, Howard Gilkinson, C. H. Haagen, R. W. Lembke, L. A. Mallory, Harry Mason, G. P. Moore, Wilbur Moore, H. M. Moser, E. J. O'Brien, Charles Pedrey, F. C. Shoup, C. H. Talley, A. G. Wesman

interview" was administered in a few moments preceding the more general classification interview. The method was adapted from speech proficiency interviews in common use in colleges. Over Navy telephones the prospective sailors pronounced numbers, repeated commands and gave an extemporaneous description of a picture. All common American speech sounds were elicited during the interview. The speech interviewer, hearing the sailor's voice over the telephones, rated him on a seven-point scale. These ratings were furnished to the classification interviewer for use in assignment.

The speech interview was never validated. Research was limited to studies of its internal characteristics and the training of personnel to administer it.[28] After training in administration, different classification interviewers correlated .80 in their ratings of speech intelligibility of typical Navy personnel. Various studies showed that the rating was not closely related to sea experience, to prior use of battle telephones, or

[28] John C. Snidecor, *A Preliminary Study of the Abilities of Rated Men to Judge Speaking Performance.* OSRD Report 1823. August 1943. Anonymous, *A Study in Training Classification Petty Officers to Select Telephone Talkers.* OSRD Report 1931. November 1943. Psychological Corporation. Washington, D.C., Applied Psychology Panel, NDRC.

to education (for Navy enlisted personnel). And the rating proved to be not closely related to scores on tests of the Psycho-Acoustic Laboratory for ability to serve as a telephone "listener."

The speech interview method was used during 1943 and early 1944 at a number of Navy centers for the classification of men for shipboard duty. It was adopted as a standard classification procedure by the Bureau of Naval Personnel on April 15, 1944.

The program of training telephone talkers began by an evaluation of a considerable variety of training methods.[29] First, Navy petty officer instructors were themselves trained in use of the methods. The instructors then trained the enlisted men who served as subjects in experiments. The efficiency of the various methods of training was tested by means of intelligibility tests administered before and after training. In the intelligibility tests, lists of words were read over the telephone to a panel of listeners. The score for a man was the per cent of words correctly understood by the panel. The members of the panel were the experimental subjects and rotated as the speakers. The results of the experiments were incorporated in outlines of training courses for Navy instructors.

Following the experiments the project staff cooperated with the Training Aids Division of the Bureau of Naval Personnel and the Interior Control Board of the Navy in preparing a Fleet Telephone Talker's Manual, or guide for personnel using battle telephones. A Supplement to the Manual, prepared by the project, included general lesson plans, special drills on digits and on communications procedure for docking, damage control, gunnery, etc. These materials were issued by the

[29] John C. Snidecor, Louis A. Mallory, and Edward L. Hearsey, *Methods of Training Telephone Talkers for Increased Intelligibility*. OSRD Report 3178. January 28, 1944. Psychological Corporation. Washington, D.C., Applied Psychology Panel, NDRC.

George W. Hibbitt and Louis A. Mallory, *Experimental Investigation of a Course for Telephone Talkers*. OSRD Report 3863. July 4, 1944. Psychological Corporation. Washington, D.C., Applied Psychology Panel, NDRC.

| I URGENT BA49 ⊞ 49 ⊞ URGENT Ø933 ⊞ Ø933 |

NEWLEAD SANFRANCISCO TOKYO RADIO SAID EARLY TODAY

THAT JAPAN IN NAME HIROHITO READY ACCEPT TERMS

ALLIED POTSDAM SURRENDER DECLARATION . BROADCAST

IN MORSE TELEGRAPHIC CODE HALTED MYSTERIOUSLY

WITHOUT COMPLETING WHAT PURPORTEDLY WAS

OFFICIAL ANNOUNCEMENT . TWO HOURS LATER

TOKYO STILL NOT RETURNED AIR .

PLATE 3. The Morse-Code Actuated Printer and a sample of a recording of a dio message made by it.

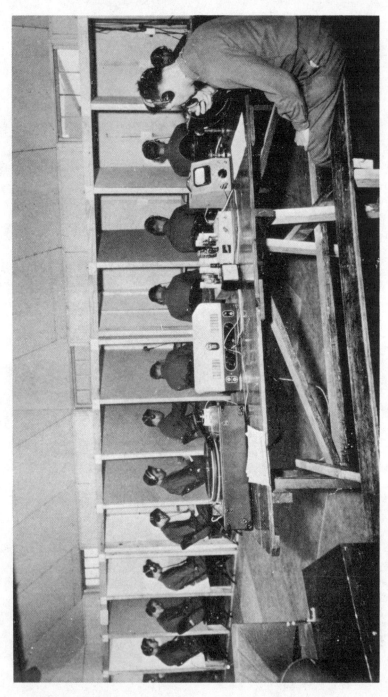

PLATE 4. Arrangement for training Army Air Forces personnel in voice communications. The instructor (foreground) communicates with the class (background) over standard communications equipment. Airplane two noise at a level of

Commander in Chief, U.S. Fleet and the Bureau of Naval Personnel to all ships and stations.

The project continued to assist the Navy through development of phonograph recordings for telephone talker training, through participation with Divisions 6 and 17 of NDRC in a program of training in submarine communications and through advisory service on courses for telephone talkers. L. A. Mallory had charge of these Panel activities which were extended to Navy West Coast operational training centers and the Hawaiian Islands.

Training in Voice Communications in the Army Air Forces

At first the Panel viewed the improvement of voice communication in the Army Air Forces as a relatively simple matter of application of existing knowledge to the preparation of a training course. A staff of psychologists and speech experts, in cooperation with Air Forces instructors, developed a training course in the use of the airplane interphone system. Fortunately the value of the course could be tested. Student pilots trained on the course were compared for intelligibility with untrained student pilots. The trained men were no better than the untrained men. The course was of no value.[30]

The result stimulated extensive research on operating procedures for airborne communications. The best methods of using the equipment, the best loudness, articulation, rate and pitch of voice, and the best forms for messages were investigated. The results were incorporated in a series of validated training courses for various kinds of aircrew and in Air Forces directives for standard operational procedures.

The basic experimental method was that of the intelligibility test described above. Developed years before the war for industrial research and during the war for Stevens' re-

[30] John W. Black, *Final Report in Summary of Work on Voice Communication.* OSRD Report 5568. September 11, 1945. Psychological Corporation. Washington, D.C., Applied Psychology Panel, NDRC.

search at the Psycho-Acoustic Laboratory, the intelligibility tests were adapted to mass use on Air Forces personnel at the Waco Laboratory by C. H. Haagen. Forty-eight lists of 24 words each were constructed. The lists were composed of common one- and two-syllable words. From these, after refinement and standardization on 50 speakers and 450 listeners, 24 lists of 12 words each were built. Scores on the final lists had a reliability of .68 for a panel of four listeners, .83 for a panel of seven and a (Spearman-Brown) predicted reliability of .91 for 20 listeners.[31] Different conditions and manners of use of the telephone or radio were then tested in terms of the per cent of words correctly understood by student-pilot listeners as one of their number spoke a list. High-level airplane-type noise filled the room during the experiments.[32]

In addition to the research by use of intelligibility tests, airborne communications were recorded as a guide to research and as a rough follow-up on results. In a study of the use of message forms, for example, communications during 40 bomber-training missions were analyzed from recordings. There were 2,377 calls. The receiver of the message failed to acknowledge the call in 35 per cent of the cases. His failure to acknowledge left the calling party in ignorance of whether or

[31] C. Hess Haagen, *Intelligibility Measurement: Techniques and Procedures Used by the Voice Communication Laboratory*. OSRD Report 3748. May 1944. *Intelligibility Measurement: Twelve-Word Tests*. OSRD Report 5414. August 4, 1945. Psychological Corporation. Washington, D.C., Applied Psychology Panel, NDRC.

Just before the end of the war Haagen converted the intelligibility tests to multiple-choice form. A sample list included these words: fog, dashboard, cold, flight, headwind, roll, missile, course, binding, practice, socket, impulse. For 169 untrained speakers, the mean score on such lists was 50.0, S.D. was 12.0, split-half reliability (corrected) for individual speakers was .86 to .94, comparability in intelligibility values of test items at different training centers was .86 to .92, and there were no significant differences by L and F tests between different lists. C. Hess Haagen, *Intelligibility Measurement: Twenty-four Word Multiple-Choice Tests*. OSRD Report 5567. September 11, 1945. Psychological Corporation. Washington, D.C., Applied Psychology Panel, NDRC.

[32] James F. Curtis, *Report on Training Studies in Voice Communication: II. The Use of Noise in a Training Program*. OSRD Report 4261. October 18, 1944. Psychological Corporation. Washington, D.C., Applied Psychology Panel, NDRC.

not the call had been heard. In a considerable number of the remaining calls the receiving party failed to understand and requested repetition of the message. Requests for repetition of the calls were made as follows:

"Say again" (standard form of the request)	16%
"Repeat"	18%
"What?" or "Huh?"	50%
Miscellaneous ("I don't understand," etc.)	16%

Such data showed the need for standardization of procedures. Standardized message forms place a minimum demand on understanding when only part of a message gets through the ever-present masking noises.

Recordings of airborne communications also guided the way to improved identification signals. The term Engineer, for example, was too often confused with Bombardier and so was replaced by Crew Chief for communication purposes.

Typical of the project research on operating procedures were its studies of voice factors. The optimal loudness of voice was subjected to investigation because directives and opinions differed. One training manual recommended that the aviation cadet should "forget the surrounding noise and imagine that he was talking directly into the ears of the listener." A second common recommendation was for the speaker to use a "conversational" level of loudness. A third suggested a "normal" or "customary" loudness level. Almost universally the student was admonished not to shout into the microphone. Nevertheless, simple physical and psychological principles suggested that the highest signal-to-noise ratio should be achieved within the limits of excessive overloading of the equipment. Hence the following type of experiment was conducted on a variety of microphones and headsets.[33]

[33] James F. Curtis, *Studies of Voice Factors Affecting the Intelligibility of Voice Communication in Noise: The Relation between Loudness of Voice and the Intelligibility of Airplane Interphone Communication.* OSRD Report 3313. February 1944. Psychological Corporation. Washington, D.C., Applied Psychology Panel, NDRC.

C. Horton Talley *et al., Report on Voice Loudness: Over Aircraft Radios*

Groups of 22 to 26 pilots were assigned from student pilot groups at Waco. Two were chosen at random to serve as speakers and the remainder served as listeners. They had some experience in communications in noise but were not fully trained. The speakers read eight 24-word lists approximately equal in difficulty and audibility. Both listeners and speakers used their communications equipment in a room filled with noise whose spectrum was comparable to that found in Army planes and whose loudness level of 108 to 110 db was also typical. The speakers systematically varied the loudness of their voices, using the maximum deflection of a meter across the microphone circuit as a means of monitoring the loudness. Four loudnesses were used corresponding to: (1) a conversational level; (2) somewhat louder, comparable to the level used in addressing an audience of 40 to 50 people; (3) the loudest level one can produce without noticeable effort to shout; and (4) shouting at the maximum level one can consistently produce. In Table 23 are shown the intelligibility scores which resulted for various microphones and headsets. Each entry gives the per cent of all words understood for speech at a particular loudness level.[34]

Under all conditions, with all types of equipment studied, either the third or fourth loudness level was the most intelligible. With some communications sets the effect was small, with others it was large, but the trend was consistent throughout. The results confirmed those obtained under S. S. Stephens of Division 17, NDRC. The results contradicted the advice to speak in a "normal" or "conversational" manner that had often been given to aircrew cadets. The instruction was revised accordingly.

Similar experiments gave the following chief results and conclusions:

and Microphones. OSRD Report 4290. October 27, 1944. Psychological Corporation. Washington, D.C., Applied Psychology Panel, NDRC.

[34] The words used in the intelligibility lists for different equipments were not equated, so results on two devices should not be taken as indicating the superiority or inferiority of one of them.

TABLE 23. The intelligibility of words spoken at four degrees of loudness over Army Air Forces communication equipment. In all cases the room was filled with airplane-type noise at 108 to 110 db. Each figure is the mean per cent of all words spoken under the stated conditions that were correctly understood. The meaning of the steps of loudness is defined in the text.

Equipment	Loudness			
	1	2	3	4 (loudest)
T–17 (hand-held) microphone				
HS–23 headset, MC–162 cushions	36.9	43.7	48.3	50.0
HS–23 headset, MC–162–A cushions	48.2	56.4	58.2	56.0
HS–33 headset, MC–162–A cushions	52.1	58.6	61.0	61.3
T–17–B (hand-held) microphone	25.8	68.4	71.4	67.2
ANB–M–C1 mask microphone	73.0	72.3	73.6	71.2
T–30–S (throat) microphone				
HS–23 headset, MC–162 cushions	18.2	31.4	33.0	30.0
HS–23 headset, MC–162–A cushions	35.1	37.7	37.9	31.8
HS–33 headset, MC–162–A cushions	45.3	47.6	49.5	40.6
SCR–183 Radio	37.8	43.4	44.3	43.6
SCR–274–N Radio	55.1	60.8	65.3	69.3
SCR–522 Radio	59.5	65.7	68.9	66.3

1. The most important factor in the use of the voice itself is loudness. In order to secure maximum intelligibility, the speaker should talk as loudly as he can without obvious strain. He should "just not shout." This is speech which produces a good loud side tone in the speaker's earphones.

2. The second most important psychological factor in voice communications is articulation.[35] Stressing final consonants and increasing the precision of articulation improves intelligibility. Even one hour of instruction given in the presence of airplane-type noise produces enough improvement in articulation to increase intelligibility significantly. Regional differences in speech are of little importance in the Army Air Forces.

[35] Harry M. Mason, *Training Studies in Voice Communication: III. Effects of Training in Articulation.* OSRD Report 5461. August 20, 1945. Psychological Corporation. Washington, D.C., Applied Psychology Panel, NDRC.

3. Rate of speech and pitch of voice have been overemphasized as significant factors for typical Army Air Forces personnel and equipment; they are not ordinarily of enough practical importance to warrant emphasis upon them in general training courses.[36]

4. Message forms and message content should be standardized, and each type of message should be as unique as possible. Message forms for interphone, radio, and special blind-landing (GCA) systems were developed by the project and incorporated in standard Army Air Forces doctrine.

5. Words vary in intelligibility.[37] Lists of words of known and widely varying intelligibility were prepared and turned over to the Army Air Forces.

6. In using the T–17 (hand-held) microphone in noise, maximum intelligibility can be secured by holding the microphones lightly touching the speaker's lips and parallel to the plane of the face.[38]

7. The T–30 and T–30–S (throat) microphones should be worn on or slightly above the Adam's apple and not below it.[39]

8. Intelligibility over the interphone is greater when the gain control is fixed at a high level than when listeners are free to adjust the volume to the level they prefer or consider most

[36] I. P. Brackett, C. Horton Talley, and Harry M. Mason, *Studies of Voice Factors Affecting the Intelligibility of Voice Communication in Noise: II. Pitch.* OSRD Report 5413. August 4, 1945. Psychological Corporation. Washington, D.C., Applied Psychology Panel, NDRC.

[37] Harry M. Mason, *Phonetic Characteristics of Words as Related to Their Intelligibility in Aircraft-Type Noise.* OSRD Report 4681. February 10, 1945. *Phonetic Characteristics Related to Intelligibility of Words in Noise: Sounds Correctly Understood in Misunderstood Words.* OSRD Report 5174. June 6, 1945. Psychological Corporation. Washington, D.C., Applied Psychology Panel, NDRC.

[38] Anonymous, *The Relative Intelligibility of Typical Methods of Holding the T–17 Microphone for Communication in Noise.* OSRD Report 3505. May 1944. Psychological Corporation. Washington, D.C., Applied Psychology Panel, NDRC.

[39] C. Horton Talley, James F. Curtis, and C. Hess Haagen, *Report on Microphone Position: T–30–S and T–17.* OSRD Report 4260. October 18, 1944. Psychological Corporation. Washington, D.C., Applied Psychology Panel, NDRC.

intelligible. Gain control should remain inoperative on the interphone.

In general these results were similar to those obtained at the Psycho-Acoustic Laboratory on laboratory personnel.

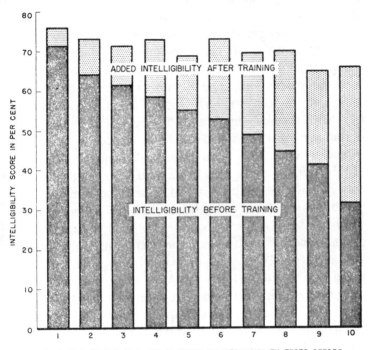

Figure 10. The effect of four hours of training on intelligibility of voice communications over an interphone system in airplane-type noise. The effect of training is to increase intelligibility; the men who were originally poor in intelligibility improved most.

The only remaining problem was to insure that actual use of the best procedures would follow the issuance of Army Air Forces directives that they be used. The principles of communication were therefore incorporated in experimental training courses which, in their turn, were evaluated by intelligibility tests. The optimal duration of such courses, the ability of Army instructors to give them in training camps, and the duration of the effects of the training, as well as the simple fact of validity, were submitted to trial.

In Figure 10 are shown the results of four hours of experimental training for one class of student pilots.[40] The improvement of 141 student pilots is shown. The group was subdivided into ten groups of decreasing original ability on the intelligibility tests. The black portions of the columns represent per cent intelligibility scores before training; the dotted portions give the same type of scores after training. All men improved, the men who were poor to begin with improving the most. Training significantly increased the intelligibility of communications and made it possible to operate with an assurance of a nearly standard rate of understanding. The value of this training course, which was based on experimentation, may be contrasted with the lack of value of the training course which was mentioned at the beginning of this discussion and which was based only on *a priori* reasoning.

The results of the training experiments were incorporated in some fifteen Training Manuals and Information Files prepared by the project for Army Air Forces. Equipment was designed to aid in instruction. A typical installation is shown in Plate 4. Separate materials adapted to each kind of personnel in the aircrew were issued. At the request of the Army Air Forces and the Signal Corps the project sent its members to various Army schools to insure proper use of its manuals. Don Lewis organized and led many of these applications by project personnel. This work was extended to the Southwest Pacific Theater by Lewis, H. M. Moser, and G. P. Moore.

The importance attached to the project's success in improving voice communication training was indicated in a

[40] James F. Curtis, *Report on Training Studies in Voice Communication: I. Can Intelligibility of Voice Communication Be Increased by Training in Voice Technique?* OSRD Report 3862. July 5, 1944. Psychological Corporation. Washington, D.C., Applied Psychology Panel, NDRC. Harry M. Mason, *Indoctrination for Voice Communication at High Altitude.* OSRD Report 5307. July 4, 1945. Psychological Corporation. Washington, D.C., Applied Psychology Panel, NDRC.

letter from Brig. Gen. F. C. Meade, USA, Director, Plans and Operations Division, Office of the Chief Signal Officer. General Meade wrote:[41]

"The Voice Communication Laboratory, located at Waco Army Air Field, and operated under NDRC Project SC–67, has developed training methods by means of which intelligibility over the interphone and radio telephone may be increased on the average by as much as 25 per cent. . . . It should be indicated that an average increase of 25 per cent in intelligibility is greater than the increase that has been obtained in recent months through costly changes in equipment."

41 Reference SPSOO 334, June 7, 1944, 3rd Ind., June 26, 1944.

CHAPTER 6

CLASSIFICATION, TRAINING, AND EQUIPMENT IN THE CONTROL OF GUNFIRE

WITH the development of military technology there has been a steady increase in the accuracy, speed, and power of gunfire. Mechanization has extended the gunner's range of information about the enemy and the fineness of his control over the guns. At the same time it has permitted the use of heavier and heavier weapons at greater and greater rates of fire.[1]

For the heavier modern weapons, and increasingly for the lighter weapons as well, the gunner sights the enemy through a complex gunsight. The simpler forms of gunsight have been replaced by complex optical, mechanical, and electronic devices. These provide information about targets by means of telescopes, pointers on dials, or spots of light on radar oscilloscope tubes.

The gunner reacts to the target or its representation by manipulating his gunsight. He adjusts handcontrol knobs, handwheels, or handlebars. These move the target relative to a reference scale, known as the reticle, within the gunsight. The gunner's task is to line up the target and reticle. This establishes the line of sight from gunsight to target.

When the target and gunner are moving relative to one another the line of sight changes and must be continuously established. This operation is known as "tracking" the target. Sometimes a single gunner tracks the target, but often the duty is divided between two men, each of whom is responsible for tracking the target in a single component. In some cases the third component of target position, range, must be de-

---

[1] This chapter and the next are based on *STR* I, Chapter 8, by W. E. Kappauf; *STR* II, Chapter 3, by W. C. Biel and W. E. Kappauf; *STR* II, Chapters 18, 19, 22, 24, and 25 by W. E. Kappauf; and *STR* II, Chapters 5 and 20, by C. W. Bray.

termined and fed into the system, either by the single gunner or by some other member of the firecontrol team.

From the changes in target position the course of the target and its rate of movement on the course are estimated. These quantities and the target's present position determine the predicted or future position of the target. Future position must be known in order to take account of the time of flight of the bullet or shell. In a gunnery system which includes firecontrol mechanisms, future position is predicted automatically or semi-automatically by a calculating device known as the computer. Computed future position is then transmitted to the guns which are aimed and fired with a minimum of human intervention.

The mechanization of gunnery and firecontrol has radically changed the task of the gunner. The nature of the change may be thrown into relief by considering the most complex human task in gunnery as studied by the Panel, the task of the gunner in a B–29 airplane. For contrast the duties of the aerial gunner in the early days of the war will be described first.

In the early days of World War II the combat job of an aerial gunner operating the flexible[2] machine guns of a bomber against enemy fighters was difficult. The gunner was generally squeezed into crowded spaces—there is little space in an airplane at best—without reference to the needs of the gunnery situation. For protection against cold and oxygen deprivation he often wore a heavy electrically-heated suit, gloves, and an oxygen mask. For long hours, sometimes isolated from his comrades, the gunner rode through enemy territory keeping constant watch for an attacker.

When an enemy plane was seen the gunner had to recognize its type. He had to note the zone in the space around his own

---

[2] A flexible gun is one which can be fired in any direction relative to the path of the gunner's own airplane. Flexible gunnery is to be contrasted with fixed gunnery, in which the guns can be brought to bear only by aiming the plane as a whole at the target. With a single exception the Panel studies of airborne gunnery were on flexible gunnery.

plane from which the enemy was approaching. When, as was commonly true, there were no reference points in the surrounding visual space, determination of the zone was difficult. If he was wise and conserved his limited supply of ammunition, the gunner waited until the enemy came within range of his own guns and until he was sure his plane was under attack. Then, in the space of a few seconds, he had to estimate how the enemy would move relative to himself during the time of flight of a bullet, aim his guns accordingly, and fire. At each pressure of the trigger the recoil of the gun moved it off the point of aim; firing had to be in intermittent bursts.

Despite its difficulty, the task of the gunner was "natural." It seemed essentially simple, or unitary, to the gunner himself. It was a task roughly comparable to that of firing a pistol from one rapidly moving car at another rapidly moving car. Disregarding the problem and the consequences of inaccuracy for the moment, the perceptual-motor task seemed clear to the gunner himself. His body movements were definitely related to the target and the various body movements required did not conflict with one another.

As the war progressed, computing machine-gun sights were developed. These eliminated much of the guesswork, particularly in the matter of computation of lead, and secured increased accuracy of fire. At the same time they complicated the gunner's task in other respects.

A diagram of one of the B–29 gunsights is shown in Figure 11. Near the top of the sight is a forehead rest which the gunner used as an aid in stabilizing the sight. Just beneath is the optical system through which the gunner viewed the target. The ring of dots and the central spot made an illuminated reticle. The fluted handwheel at the extreme left served as the azimuth-elevation tracking control with which the central reticle dot was held on the target. To track in azimuth, the gunner moved the sight as a whole around its verticle axis. To track in elevation he rotated the handwheel around its own axis.

The handwheel shown at the extreme right of the sight was used to determine the range of the target. Rotation of this range control about its own axis changed the diameter of the reticle circle. For a head-on attack the gunner tried to keep the target wing-tips just framed in the circle of dots. For other attacks the diameter of the circle of dots was kept at the estimated width from wing-tip to wing-tip. Before the attack began, the relation of circle size to range was adjusted for the estimated size of the enemy plane; thus, when target identification and framing were correct, the size of the reticle circle gave the range of the target.

Although the two hands had separate tasks it was necessary that they work together to provide stability to the sight. This was a particularly difficult problem for the right hand.

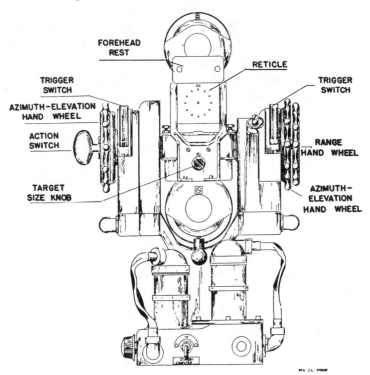

Figure 11. The B–29 Pedestal Gunsight. See text for explanation.

The inner one of the two handwheels at the right in Figure 11 was a secondary azimuth-elevation control. It was lightly held in the fingertips of the right hand. As the gunner changed range on the target his fingers slid over this inner handwheel, holding it just firmly enough to stabilize the sight in azimuth and elevation, yet not so firmly as to interfere with ranging or tracking.

The elevation and range controls were linked internally. In consequence the movements of the range handwheel had to be made relative to movements of the sight in elevation. Depending upon the direction and rate of elevation changes relative to range rate, the rate of hand movement in ranging might differ for the same target range rate.

It was a difficult problem for the gunner to learn to tie together the asymmetrical movements of the two hands in response to the two quite separate aspects of the target stimulus, position, and range change. Coordination of these unit tasks was as difficult as patting the head with one hand and rubbing the stomach with the other—with the rhythm of each hand determined by separate external stimuli.

To the complexities of tracking and ranging was added the duty of pressing the trigger. Two identical trigger thumb-switches, for use of either thumb in triggering, were provided. They are shown in Figure 11 next to the handwheels. The triggers were mounted on a portion of the sight which moved in azimuth tracking but not in elevation tracking. Thus each trigger moved relative to its thumb during elevation tracking and the right trigger also moved relative to the right thumb during range tracking.

The gunners were instructed to track accurately and smoothly, to range accurately, to begin firing when tracking and ranging were adequate, and to fire intermittently as long as tracking and ranging continued to be adequate.

The task of the B–29 gunner was complex; nevertheless the gunner was successful. Japanese attacks on the B–29 were fought off with relative ease. It was evident, however, that simplification of the gunner's task held possibilities for even greater success.

Turning now to firecontrol systems in general, the mechanization of gunnery produced the following three psychological problems:

1. The relation of gunner to target and guns was indirect in the mechanized system. In many systems the target could not be seen directly. The representation of a target by a dial pointer or an oscilloscope blip did not necessarily look like the target. In such presentations the actual movements of the target might be changed in apparent magnitude, direction and rate. Likewise, the gunner was separated from his guns. He operated them through gear trains, computing devices, or motors. These often modified the gunner's movements in direction, amplitude, and rate.

2. The number of unit tasks was increased. Frequently several gunners had to cooperate. Often the individual gunner had several relatively distinct and separate tasks. Successful performance required that all units be carried out adequately at the same time. Failure resulted from deficiencies in any single unit.

3. Mechanization required that delicate complex equipment be maintained and calibrated. This task was usually the primary responsibility of a specialized maintenance man, but the gunner had at least a minimum of operational maintenance and calibration to perform for himself. Success in these relations to materiel was absolutely essential if hits were to be scored on the target. It was an easy matter to correct an error in a gunsight when one observed directly that a projectile fell short or over, to right or to left. When a projectile fell out of sight and when many unit activities determined the precise point at which a projectile dropped, the correction of errors was difficult.

The indirect relations, the number of units, and the problems of maintenance and calibration combined to prevent the individual gunner and his officers from knowing whether his duties were well or poorly done. Knowledge of the results of the gunner's actions was distorted, delayed, or even prevented altogether.

It is an elementary psychological principle that the best performance cannot be attained without knowledge of results. When knowledge of results is lacking, good and poor performance cannot be differentiated either by the man or by his instructors. Good performance cannot be praised, poor performance cannot be criticized, except in a non-informative way. Competition between men, and even self-competition, cannot be encouraged. During World War II these factors created successive difficulties with gunnery and firecontrol personnel and led to a series of requests for research by the Panel.

In the summer of 1942 a request was received to study the selection and training of Army heightfinder operators. Research on the latter project spread into the Navy and in 1943 resulted in a Panel project on the selection and training of rangefinder and firecontrol radar operators. Later in 1943 the work extended into studies of operator performance on all types of Army antiaircraft equipment. At the end of 1943 the Panel was asked to study psychological problems in the design and operation of new Navy lead-computing gunsights and directors. In May 1944 a project on field artillery gunsights was undertaken. And in July 1944 the Panel began its research on B–29 airplane gunnery.

The origins and general nature of the work of each of these projects will be described in this chapter. In the next chapter the basic methods and the actual results of a few of the research studies from the gunnery and firecontrol projects will be considered.

STEREOSCOPIC HEIGHTFINDER AND RANGEFINDER OPERATORS

In the spring of 1941, Division 7, the firecontrol division of the National Defense Research Committee, undertook to improve the use of the stereoscopic rangefinder. This instrument is a binocular telescope which provides information about the distance to a target. The stereoscopic rangefinder provides each eye of an observer with a separate, magnified view of the target. The two views differ from one another by an amount which is a function of the range of the target and of the distance between the two end mirrors through which the two views of the target are taken. Within each telescope is a set of reference marks composing the reticle. The operator must fuse the two views of the target and the two sets of reticle marks into single images of one target and one reticle just as he fuses any pair of images from his two eyes into a single image. When fusion occurs the operator sees the target and reticle in relative depth. The target appears closer or farther away than the reticle. By manipulating the range knob, which affects a set of prisms in the instrument, the operator makes the target appear to be at the same distance as the reticle. At this point the position of the prisms corresponds to the angle of convergence of the two lines of sight on the target. This angle is directly related to the range of the target and is shown in terms of range on the range scale. Essentially the operator triangulates on the target, using the distance between the two end mirrors as the base of a rangefinding triangle.

The stereoscopic rangefinder was in common use in the Navy. The Army used a modification of it, known as the heightfinder. The height of a target above such an instrument is a simple function of range and the angle of elevation of the instrument. The job of the operators, in so far as Panel studies were concerned, was the same whether height or range was measured.

The accuracy of measurement of height or range depended upon the perceptual ability and skill of the operator and upon the condition of his instrument. If the operator was skillful and if his instrument was in good condition, accuracy should have increased as the operator's stereoscopic acuity increased and as base length and magnification increased. Actually these theoretical relations did not hold, even for skillful operators and instruments in good condition. It was this difference between expected and obtained results which led Division 7 to an extensive program of research on the improvement of the accuracy of heightfinder operators.[3]

Division 7 made many fundamental contributions to knowledge of the heightfinder and its operator. It showed that errors in height readings resulted from instability of the instrument, inaccuracies of instrument adjustments, inaccuracies in calibrating procedures, and variability in stereoscopic judgment. It developed and proved the value of many corrective devices and procedures for these sources of error.

When the research of the Applied Psychology Panel began in 1942, Division 7 had completed many of its studies of the stereoscopic heightfinder. Two problems relevant to personnel were still under study. The validity of selection tests for the heightfinder operator had not been completely established, and the training of operators was relatively ineffective in securing use of the Division's improved operating procedures. The Division requested the Panel to take over the field work on selection and training.

At first the Panel was inclined to reject the request. Radar was developing as a tool for rangefinding; obsolescence of the stereoscopic method was a possibility. The conditions for validation of the Division's visual aptitude tests were likely to be unsatisfactory. The tests were already in use; to obtain adequate proof of their validity would be difficult, and even positive results might add little to the winning of the war.

On the other hand the radar gear available at the time was not available in quantity. Its upkeep was difficult and it was not always superior to the stereoscopic method in accuracy (in actuality the rangefinder proved to be the standard instrument to the end of the war in some critical Navy situations). In addition, revision of the training program for the heightfinder operator was quite desirable. Adequate training would

---

[3] The program of Division 7 was under the supervision of S. W. Fernberger in so far as its psychological aspects were concerned.

insure that at least one good method of ranging was available for service use. Finally, the Panel felt that the National Defense Research Committee had an obligation to try to validate any selection tests which it might recommend.

After arranging with the Army's Antiaircraft Command to improve research conditions, the Committee accepted the request of Division 7. From the resulting project on selection and training of heightfinder operators came the whole series of remaining Panel projects dealing with classification, training, and equipment in the field of firecontrol.

C. H. Graham was placed in charge of the Panel's work. He was assisted by W. J. Brogden and the staff shown in Table 24. W. E. Kappauf was chiefly responsible for the ac-

TABLE 24. The research staff of the Panel project on heightfinder operators.

Contractor: Brown University
Contractor's Technical Representative: C. H. Graham
Project Director: W. J. Brogden
Staff: J. K. Adams, W. C. Biel, W. F. Dearborn, D. G. Ellson, H. M. Fisher, H. M. Fowler, H. A. Imus, W. E. Kappauf, E. B. Knauft, B. McMillan, F. A. Mote, M. W. S. Swan

tual experimental work and the applications of the results.[4]

The Panel continued the studies begun by Division 7 on the interrelations between classification, training, and equipment. The procedures for operating the heightfinder and the possibility of measuring the skill of the heightfinder operator in use of the procedures were the threads which tied these subjects together. When adequate operating procedures were unknown, a program of selection and training would have little point. Instrumental errors were so great that the quality of the operators made little or no difference. Nevertheless, even when improved operating procedures were known they were often not utilized. Under these circumstances only a training

[4] The final summarizing report of this project is not available to the public.

program could secure their use. Until improved operating procedures were applied and until operator skill could be measured it was impossible to validate selection tests.

In research on operating procedures the Panel concentrated on field maintenance and calibration of the instrument. Previously the procedures used by the operators in training had included or even depended upon conditions which held good only in school. Procedures suitable for use in the field were developed and substituted. The procedures were described in step-by-step outlines and the men were trained in their use. Experimental trial with a good criterion proved the validity of the procedures and of the training methods as well.

The major work in the field of training was to develop a complete manual for the operator and numerous supplementary lesson plans, suggestions and other aids for the instructor. The manual and associated aids described proven methods for evaluating operator proficiency in many subordinate phases of the task and prescribed optimal durations of training based on learning curves. The use of Division 7's synthetic training devices was described in the manual. The manual and associated aids helped to secure a complete revision of heightfinder and rangefinder operator training in the Army and Navy.[5]

The project validated the visual tests for stereoscopic heightfinder operators (see Chapter 7) and gave assistance to the Army and Navy in setting up visual test centers to insure adherence to the visual standards which resulted in both services.[6]

As the Panel project on the heightfinder operator developed, the Bureau of Ordnance of the Navy became interested in the application of the results to Navy rangefinder personnel. The Bureau requested the Panel to prepare manuscript for ordnance pamphlets on the rangefinder and on synthetic trainers

[5] The manual and associated aids were prepared under W. E. Kappauf. Material from Kappauf's manual appeared in War Department Manuals FM 4-142, TM 44-250, TM 9-623, and TM 9-624.

[6] H. A. Imus was in charge of the Panel's assistance to the Army and Navy in setting up visual test centers.

for the rangefinder. J. L. Kennedy of the Panel staff was assigned to the job. Kennedy's first problem was to become acquainted with Navy operating procedures and firecontrol equipment and to adapt the facts learned in the Army height-finder project to them.

In order to familiarize Kennedy with Navy firecontrol problems he was assigned to part-time duty at the Advanced Fire Control Schools, Washington Navy Yard, where Lt. Comdr. (later Comdr.) A. L. Shepherd, USN, was officer in charge. These schools included instruction on the rangefinder. Working together, Kennedy and Shepherd prepared psychological literature for the ordnance pamphlets on the selection and training of rangefinder operators.

At the time, Navy training of rangefinder operators at the Advanced Fire Control School was more theoretical and limited to materiel than seemed desirable in view of the Panel's experience in the Army. There was a need for increased emphasis on operations and on calibration procedure. Division 7 had shown that stereoscopic operators required several months of systematic operational training before a peak of performance was reached. The Navy school gave less time than this, and its training in operations was relatively brief and unsystematic.

Increasing familiarity with the Panel's Army research led Shepherd, with Kennedy's help, to work up a new program for the training of Navy rangefinder personnel.[7] As a result, in August 1943 a new school for rangefinder and firecontrol radar operators was organized at Fort Lauderdale, Florida. Its provisional curriculum included adaptations from the Panel curriculum for the Army heightfinder school. At the same time the Panel was asked to undertake research on the Selection and Training of Rangefinder and Radar Operators. As the Panel project on Army heightfinder operators was then terminating, it seemed desirable to carry over all the project's

---

[7] D. G. Ellson assisted Kennedy and Shepherd in the later phases of this work.

knowledge to the Navy. The request was accepted and the staff of the heightfinder project moved to Fort Lauderdale with the opening of the Navy school. W. J. Brogden was placed in charge of research on rangefinder and radar operators.[8] He was assisted by the staff listed in Table 25.

TABLE 25. The research staff of the Panel project on rangefinder and fire-control radar operators.

Contractor: University of Wisconsin
Contractor's Technical Representative: W. J. Brogden
Project Director: D. G. Ellson
Staff: J. K. Adams, D. C. Beier, H. M. Fowler, F. E. Gray, H. A. Imus, W. E. Kappauf, E. B. Knauft, F. A. Mote, S. D. S. Spragg, M. W. S. Swan, H. A. Taylor

The Panel developed and assisted the Navy in applying the training methods of the heightfinder operator project. It prepared lesson plans for rangefinder officers, developed a new criterion for the measurement of rangefinder operator performance, developed achievement and proficiency tests for the operators of other firecontrol gear as well as for the rangefinder operator, designed and evaluated synthetic trainers for antiaircraft personnel, and re-validated the Army heightfinder selection tests for use with the Navy rangefinder operator.

NAVY FIRECONTROL MANUALS

In the meantime Commander Shepherd had been made responsible for a whole program of revision of materiel manuals for firecontrol gear. Shepherd sought Panel assistance.

The Panel was then considering a suggestion that a civilian center for the preparation of technical manuals might be created under Panel auspices. The problem was to make the complexities of the new materiel clear to the thousands of soldiers and sailors who had suddenly to be trained as mainte-

[8] The final summarizing report of this project is not available to the public.

nance men. A survey of the field was undertaken and one of the proposed manuals was written[9] as a trial balloon. The manual was used throughout the Navy, but the problems of preparation of technical maintenance manuals were too remote from psychology to warrant continuing Panel interest, and the Panel dropped the subject.

ARMY ANTIAIRCRAFT ARTILLERY

When the heightfinder project closed and its research moved to the Navy, the Antiaircraft Command of the Army requested the Panel to continue its research on firecontrol for antiaircraft guns in a Study of Operator Performance on All Types of Antiaircraft Equipment. L. C. Mead, with the aid of W. C. Biel and the staff listed in Table 26, carried on the work.[10]

TABLE 26. The research staff of the Panel project on personnnel of the Army antiaircraft artillery.

Contractor: Tufts College
Contractor's Technical Representatives: J. G. Beebe-Center, L. C. Mead
Project Directors: L. C. Mead, W. C. Biel
Staff: R. N. Berry, G. E. Brown, Jr., R. M. Gottsdanker, W. J. Griffiths, Jr., R. C. Hall, A. C. Hoffman, B. B. Hudson, J. H. Rapparlie, L. V. Searle, A. L. Sweet, K. S. Wagoner

---

This project undertook three types of study. First was the validation of a synthetic tracking trainer[11] originally de-

---

[9] By P. W. Johnston, under the contract with the National Academy of Sciences.

[10] The final summarizing report of this project is not available to the public. A few reports of studies of fatigue were published under Panel auspices. The fatigue studies began under Division 7, NDRC: Arthur C. Hoffman and Leonard C. Mead, *The Performance of Trained Subjects on a Complex Task of Four Hours Duration.* OSRD Report 1701. July 24, 1943. Leonard C. Mead, *Summary Report of Research and Development Work from August 1, 1942 to July 1, 1943.* OSRD Report 1592. June 30, 1943. Tufts College. Washington, D.C., Applied Psychology Panel, NDRC.

[11] Bradford B. Hudson and Lloyd V. Searle, *Description of the Tufts Tracking Trainer.* OSRD Report 3286. February 5, 1944. Tufts College. Washington, D.C., Applied Psychology Panel, NDRC.

veloped at Tufts College for Division 7; training on the trainer proved to be as effective as training on the gun director which it simulated. Second was the development of proficiency tests for antiaircraft trackers and research on the usefulness of the tests in training. Third, the project assisted the Antiaircraft Board in acceptance tests of new equipment and alternative operating procedures. In such tests psychologists were needed to assure consideration of the personnel factor.

FIELD ARTILLERY GUNSIGHTS AND PROCEDURES

In the operation of field artillery the target was frequently invisible to the officers and men at the guns. The guns usually operated in a battery of four under the direct command of the executive officer. To begin with, the four guns had to be aligned with one another and with reference to some aiming point. In opening fire on a target the guns were fired as accurately as possible, errors were corrected under the guidance of a forward observer, and fire continued. When a new target was chosen the process was repeated. An elaborate system for the transmission of intelligence was required. Typically the executive officer received information via telephone from a fire direction center which in its turn had received telephonic reports on the location of shell explosions from a forward observation post.

The occurrence of frequent and serious errors in the operation of field artillery guns had been recognized for years. In regular peacetime training there was time to give full consideration to the sources of error and to select the few operators required after partial training had revealed their ability to operate with a minimum of error. Careful instruction, repeated drill, prolonged training, and ultimate selection of a small number of men practically eliminated some types of error. In the war emergency, the accelerated training program did not permit successful use of these common sense empirical procedures.

Large errors in field artillery fire are always serious since they involve the possibility of hitting one's own troops. A commonly recognized and publicized error is the 100 mil error. When an error of 100 mils in deflection occurs in shooting at a target which, for example, is 5,000 yards distant, the shot lands roughly 500 yards to one side or the other of the intended target. If the opposing armies are close together, if the lines are irregular or in motion, such an error may result in fire on one's own troops.

In May 1944 Army Ground Forces requested the Applied Psychology Panel to cooperate with the Armored Medical Research Laboratory (abb.: AMRL) of Fort Knox, Kentucky, on a study of the sources of the 100 mil and other errors in field artillery fire. J. P. Nafe and the staff listed in Table 27 represented the Panel.[12]

TABLE 27. The research staff of the Panel project on field artillery sights.

Contractor: Tufts College
Contractor's Technical Representatives: J. G. Beebe-Center, L. C. Mead
Project Director: J. P. Nafe
Staff: R. N. Berry, R. H. Brown, L. V. Searle, K. S. Wagoner

A joint preliminary analysis of field artillery practices was made by AMRL and the Panel. It developed that errors occurred at each of the following points:

1. In survey: errors in measuring angles, measuring distances, and in computing coordinates.

2. At the battery: errors in aligning the guns of the battery, confusion in commands, and failure in communications while laying the guns; improper execution of commands; and errors in use of the gunsights in setting and reading scales.

3. At the observation post: reversals of direction in re-

---

[12] Only two reports from the project are available to the public (see Chapter 7).

questing desired shifts in fire, failure to send clear reports, mis-reading of instruments.

4. In communication: errors due to similarity of sounds, confusion of four digit numbers, poor enunciation, etc.

5. At the fire direction center: errors in transmitting messages, making computations, and reading instruments.

The major sources of error were found in departure from standard operating procedures and failures to use the equipment correctly. The frequency of errors was significant and concentrated at the guns and in communications.

From the outset the two laboratory groups from AMRL and the Panel made plans for cooperative work on their problems. In general it was decided that the Panel group would investigate the nature, distribution, magnitude, and correction of errors in the use of gunsights and associated equipment, while the AMRL group would study the procedures of fire direction. Although it was possible to conceive of system improvements accomplished, for example, by a complete change in the data transmission system to the guns—a substitution of selsyn transmission for telephone transmission—work toward such changes was left as a peacetime job and was set aside in favor of developmental work which might lead to the immediate improvement of field performance with equipment not fundamentally different from that in use. It was decided that both research groups should concentrate at first on special problems associated with the reading and setting of scales on field artillery instruments.

Working together, AMRL and the Panel determined the frequency of errors in the use of field artillery sights and compared the frequency on various types of sights (see Chapter 7). They developed several pilot models of new gunsight scales which were designed to eliminate many of the errors made at the guns. In addition they developed a complete system for recording the source and magnitude of all errors in the operation of a field artillery battery (see Chapter 7), and they recommended that Army Ground Forces personnel receive training in voice communication over telephone and radio systems. As a result the Field Artillery requested the Panel to undertake a project on the training of personnel in telephone communications. The Panel rejected the request, partly because time and personnel were not available and partly as a result of the belief that further research on voice communications, at least research of the kind contemplated, was not required. Previous Panel studies (see Chapter 5) had clearly defined the training problems in voice communication for the conditions of service use and the ways of solving these problems. In the opinion of the Panel it was up to the Army to administer the results of research in all of its units. Advisory assistance without research was offered to the Field Artillery. Ultimately, a few courses in voice communication patterned after those described above in Chapter 5 were given in Army Ground Forces units by Signal Corps psychologists.

B-29 GUNNERY

In July 1944 the Army Air Forces requested the Panel to undertake a study of Psychological Factors in the Operation of Flexible Gunnery Equipment. The request was as follows:[13]

"(1) It is requested that the following project be submitted to the National Defense Research Committee: AC-94—Psychological Factors in the Operation of Flexible Gunnery Equipment. (2) The object of the project should be to conduct research and secure information necessary to establish military requirements that will insure the procurement of the types of equipment that may be most effectively operated by the average flexible gunner under combat conditions. (3) The increasing operational use of heavy bombardment aircraft has led to a corresponding increase in the defensive armament and fire power of these airplanes and has resulted in a rapid

[13] Col. J. F. Phillips, Air Forces Liaison Officer for NDRC to War Department Liaison Officer for NDRC, Subject: Submission of new project to the NDRC, July 13, 1944.

increase in the responsibility placed upon the flexible gunner and an increase in the difficulty and complexity of the gunner's task. Although rapid progress has been made in the development of turrets and firing equipment, only a limited amount of information is available concerning the ability of the gunner to use this equipment effectively against the enemy. If further improvements are to be made, and adequate protection is to be given to heavy bombers in missions over enemy territory, it is necessary that information be secured concerning the human limitations of the gunner, especially limitations in his ability to learn to use different types of equipment. It is especially important that requirements for equipment be established so that the average gunner can learn to use the equipment as rapidly and effectively as possible."

The request was accepted with the proviso that research would be limited to the problems of B–29 gunnery. The limitation was required by the inability of the Panel to locate additional research personnel. Even the qualified acceptance of the project required a reduction or termination of research on other topics. The staff was drawn almost entirely from pre-existing Panel projects.[14] W. J. Brogden, with the help of

TABLE 28. The research staff of the Panel project on B–29 gunnery.

Contractor: University of Wisconsin.
Contractor's Technical Representative: W. J. Brogden
Project Director: K. U. Smith
Staff: J. K. Adams, D. C. Beier, C. S. Bridgman, D. G. Ellson, F. E. Gray, J. E. P. Libby, L. H. Lanier, W. C. H. Prentice, R. R. Sears, R. L. Solomon, C. H. Wedell
Subcontractor: State University of Iowa
Subcontractor's Technical Representative: R. R. Sears
Staff: E. B. Knauft, C. E. Osgood, R. Phillips, D. M. Purdy

---

[14] The process of robbing Peter to pay Paul had been the Panel's method of staffing several previous projects. For some time the Panel had followed a policy of refusing new projects unless their priority was formally stated by the War or Navy Department to justify the draft deferment of research workers.

K. U. Smith and the staff listed in Table 28, conducted the work.[15]

A large field laboratory was established by the Panel at the Research Division, Laredo Army Air Field, Texas, to study B–29 gunsights and gunners. At Laredo the Research Division was charged with studies of the selection and training of aviation gunners. The officer in charge, Maj. N. Hobbs, AUS, with Maj. R. W. Russell, AUS, Lt. D. Howe, AUS, Lt. T. R. Vallance, AUS, S/Sgt. B. Wollen, and others cooperated with the Panel in making Army facilities available and in the planning and conduct of research. Personnel and equipment from the Research Division were loaned to the Panel and vice versa; members of one group represented the other at conferences. In general the division of responsibility for psychological problems of aerial gunnery followed the original directives to the two groups. The Panel project took over most of the experimental developments in the field of equipment design. The Research Division continued its responsibility for the selection and training of gunners, but the separation of functions broke down frequently by mutual consent.

The Panel project was chiefly concerned with fundamental methods for the formulation of the personnel requirements in equipment design. Formulation of general requirements was possible only if basic research and systematic studies could be completed. The staff therefore concentrated on the development of an experimental test unit in which the effects of systematic variation in equipment could be measured. The problem was to provide meaningful measures and systematic variation of conditions in a convenient yet realistic experimental framework. The solution of this problem will be described in the next chapter. The solution provided a series of by-products in the form of proficiency tests, scoring devices, and synthetic trainers.

In addition to their fundamental research the project staff

[15] Only one report from the Panel's research on B–29 gunnery is available to the general public (see Chapter 7).

at Laredo developed and evaluated several sets of new and simplified gunsight controls. They carried on a continuous advisory service to Army Air Forces and to industrial concerns on psychological needs in the construction of gunsights. New gunsights for the Army Air Forces will be better adapted to gunners as a result of the work.

The research group at Laredo was supplemented by a second group from the State University of Iowa under R. R. Sears at Smoky Hill Army Air Field, Salina, Kansas. Here the emphasis was on immediate development of proficiency tests for gunners in operational training. Scoring systems were desired to provide knowledge-of-results training and competition among gunners. They were also needed to provide a means of validating the various training devices used in operational training. Emphasis was placed on scoring methods for gun-camera film, and on scoring performance on various mock-up training devices.

THE ORGANIZATION OF PSYCHOLOGICAL RESEARCH ON FIRECONTROL

The sequence of projects in gunnery and firecontrol has been outlined in some detail because in these projects the Panel most nearly approached the coordinated integrated research program which was contemplated when the Committee on Service Personnel was created. The firecontrol projects of the Panel assisted one another by loans of personnel, by sharing results on methodology, by passing information back and forth. Results secured in one service were applied in the other. Problems of classification, training, and equipment were studied together, as they may most effectively be studied, as various aspects of the same fundamental problem, the improvement of human proficiency.

At least in one respect, however, the organization of the above-described Panel research projects on firecontrol was deficient. The timing of the work was poor throughout. In every case mentioned above it is possible to observe that while the main research results were of great practical value in the war, they were available too late to be of maximum service. In none of the projects so far described did research even begin until well after personnel difficulties had been experienced in combat. The research itself then required time. General application of results by the services usually required at least as much time as the research itself. In a war whose character changed so rapidly as World War II, time was of the essence.

In the early days of the war in the Atlantic and in Europe and, later, as war developed in the Pacific, the airplane dominated the scene. The absolute necessity and the partial failure to protect ships and ground stations from enemy air attacks was followed by successive requests, beginning in 1942 and ending in 1943, for research on naval gun crews, heightfinder operators, rangefinder and radar operators, the personnel of the whole antiaircraft battery, and one additional personnel group which is still to be described.

America entered the ground war in late 1942. During 1943 field artillery problems rose to higher and higher priority. In May 1944 a request was received for psychological studies of field artillery methods and devices.

The B–29 airplane appeared on the combat scene in 1944. The dissatisfaction of gunners with their combat equipment and the rising threat of suicide attacks were followed in July 1944 by a request to study the B–29 gunsight.

These experiences of the Panel were paralleled by the experiences of military psychologists in general (and perhaps by the experiences of a fair proportion of all scientists). Psychological research could hardly be said to be too little and too late, but it was certainly too late for maximal effectiveness.

The new devices on which high priorities were placed and on which psychological problems were acute were designed for combat. When placed in training camps or research laboratories they did not immediately contribute to the war against the German and the Jap. So production models went at first to combat and not to training or to psychological research. Under the circumstances service training was necessarily abstract. Classification procedures were oriented toward training. Research on personnel was delayed. The needs of personnel were frequently not anticipated in the new equipment.

In the organization of research, therefore, one problem was to develop a system for adequate consideration of personnel requirements in anticipation of the need. The problem was sharpened, in the case of a particular new Navy gunsight, by general misuse of the new device in combat. The gunsight was known to have great potential value, but it involved certain radical departures from old Navy sights. Experience with old gear was not very useful in operating the new. Misuse of the gunsight was general and in some instances the high priorities on scientific and production effort which rushed it to the Fleet were completely wasted; officers and men locked the gunsight out of operating condition after it was installed.

The problem was acutely felt not only by the Panel but also by Commander Shepherd of the Advanced Fire Control School, by the Readiness Division of the Office of the Commander in Chief, and by the Research Division of the Bureau of Ordnance. In the Readiness Division, Comdr. C. S. Martell, USN, was responsible for fleet training in the operation of new directors using a gunsight of this type and of the prospective improved models then on the drawing boards. In the Bureau of Ordnance, Lt. P. G. Nutting, USNR, had recently been assigned to deal with training problems.

With the assistance of J. L. Kennedy of the Panel staff, Martell, Nutting, and Shepherd developed a program to reduce personnel trouble with the successive new gunsights which were under design. For the Panel, the program centered on a project named Psychological Problems in the Design and Operation of Gyroscopic, Lead-Computing Gunsights and Directors. In the opinion of the Panel staff the organization of this project should be considered as a prototype for the organization of future research in military psychology.

LEAD-COMPUTING GUNSIGHTS AND DIRECTORS

In essence the project consisted of a psychological group which worked with materiel men on preproduction models (and later on blueprints and mock-ups) of new Navy gun directors. The psychological group studied the design of the directors in terms of personnel needs. It developed and tested modifications to the designs and standard operating procedures. It trained personnel in use of the procedures for field trials of the preproduction models and assisted in the control of personnel factors which might affect the field trials. In addition it developed suggestions for the training and selection of the personnel who would operate the device in combat. These activities occurred before production of the directors began, and results were available as the new directors came off the production line.[16]

In this process the project became a liaison center for personnel men, materiel men, and combat officers of the Navy, as well as for civilian research workers and representatives of industry. The psychologists of the project with their background in science and their interest in the human being could talk the language of all these groups and help to integrate their work.

For the Panel, W. E. Kappauf was placed in charge under the general direction of C. H. Graham. Kappauf began research with a preproduction model of a new gun director, the Mark 52, in the last few days of 1943. Shortly thereafter F. V. Taylor, A. S. Householder, and the staff listed in Table 29 were recruited. In addition a number of Navy officers were attached to the project from time to time.

The first work concentrated on the development of standard operating procedures. Kappauf, and later Taylor, developed the procedures with the complete cooperation and ac-

[16] A summary and complete bibliography of the project will be found in W. E. Kappauf, *Summary of Research on Psychological Problems in the Operation of Antiaircraft Lead Computing Sights and Directors: Final Report of Contract OEMsr–815.* OSRD Report 5425. September 29, 1945. Brown University. Washington, D.C., Applied Psychology Panel, NDRC.

TABLE 29. The research staff of the Panel project on lead-computing gunsights and directors.

Contractor: Brown University
Contractor's Technical Representatives: C. H. Graham, W. E. Kappauf
Project Directors: W. E. Kappauf, F. V. Taylor
Staff: H. P. Birmingham, T. G. Hermans, M. W. Horowitz, A. S. Householder, W. W. Lambert, H. D. Meyer
Contractor: National Academy of Sciences
Subcontractor: The University of Rochester
Subcontractor's Technical Representative: J. D. Coakley

---

tive participation of Martell, Shepherd, and Nutting. The function of the gunsight, the nature of its mechanism, its role in defense tactics, and its relation to other equipment on shipboard were studied in detail. A variety of possible operating procedures was tried out. The nature of the trials ranged from rough checks to fairly precise experimentation, usually based on time as a measure. The time required to solve various firecontrol problems and to get shells in the neighborhood of the target was critical in shipboard antiaircraft gunnery.

Within a few weeks a preliminary set of operating instructions was ready for the field trials of the director, and several crews were trained in their use. The results of the field trials gave an empirical check on the efficiency of the tentative operating procedures. The procedures were modified where necessary and a manuscript describing them was turned over to Commander Martell. In Martell's office the manuscript was revised for fleet publication. The final manuscript was ready at the time the production models came off the line. A similar program was followed for a series of new directors.

The work on the operating procedures stimulated the development of optimal training methods, of standards of proficiency, of the probable duration of practice required to meet these standards, and of other training matters. These were described in the operating instructions, which also included methods for the use of proficiency tests in the selection of personnel on shipboard.

The development of standard operating procedures permitted Kappauf and Taylor to detect certain design modifications which would facilitate operations and training. In a few cases the need for modifications in design was a result of the failure of designers of parts of the system to be informed of the characteristics of other parts. The most extreme examples were cases in which one new gunsight could only be operated by men over six feet tall and then only on a smooth sea, whereas another new gunsight would have required one man to spend much of his time moving about a circle ten feet in diameter on hands and knees. These obvious defects were born of wartime emergency designing procedures and would have been corrected by any group which first tried to use the new device. The project's corrective measure was a simple adjustable-platform method for suiting the device to the height variations of the critical members of the gun director crew. Less obvious problems were the need for counterbalances, the design of supports for the gunner which would not add leverage to his arms as he tracked the sight on a pitching deck, the best reticle pattern, a gear mechanism which moved the sight telescope through an arc adapted to the movement of the human eye in elevation tracking,[17] and a checksight and method for using it in training. On successive preproduction models of new gunsights of the gyroscopic lead-computing type these and similar changes in design were studied and in many cases installed in time to appear in the first production models.

The studies also included the evaluation of a series of gunnery trainers before production of these began. The whole program made it possible for the staff and Navy officers to prepare lesson plans and aids for instructors, achievement and proficiency tests, and curricula. It seems obvious that the project laboratory would have provided the best location for preliminary research on the development of aptitude tests had such tests been assigned priority at the time. The success of the project is indicated by the fact that its work is continuing

[17] This arc was determined by T. G. Hermans.

in the postwar period under F. V. Taylor at the Naval Research Laboratory, Washington, D.C.[18]

Essentially the project on lead-computing gunsights and directors provided the organization required to complete personnel research in anticipation of the need for it. The extra cost of this research organization in comparison with other kinds of organization was negligible in terms of the delivery date for practical results. The cost was the sacrifice for research purposes of one preproduction model of each new device and the wholehearted, intelligent cooperation of the project's liaison officers. The experience of this project will be considered again in Chapter 9 in a discussion of future research in military psychology.

[18] A comparable research unit has been established at Wright Field, Dayton, Ohio, the Army Air Forces center for development of new equipment. P. M. Fitts is in charge of the psychological group at Wright Field.

CHAPTER 7

EXPERIMENTAL METHODS AND RESULTS IN THE CONTROL OF GUNFIRE

I N the study of the human factor in the control of gunfire the basic psychological problem was always to develop accurate, reliable, and meaningful measures of the performance of personnel.[1] Whenever such measurements were obtained, experimentation with practical and theoretical values became possible. In the absence of performance measurements experimental research was impossible. Since the problem of the criterion was so basic, the various methods of measurement will be described independently of the results of the experiments. A comparison of methods will illustrate the unity of the psychological approach; a comparison of results will show the range of human problems which yield to fundamentally similar methods of investigation.

METHODS OF MEASUREMENT

Stereoscopic Heightfinder and Rangefinder Operators

In its first research on firecontrol the Panel undertook to continue the studies of Division 7, NDRC, on the stereoscopic heightfinder. The Panel project on the selection and training of heightfinder operators took over the methods of Division 7 in order to validate proposed selection tests and training methods. The basic research problem was to obtain a satisfactory criterion measure of the performance of the operator and his equipment. A criterion measure could be obtained only if the "true" height of an airplane target was known at the instants when the operator read the target height through his instrument; the height of the target had to be measured independently of the operator's measurement, and with an

---

[1] The direct sources of this chapter are described in footnote 1 of Chapter 6.

accuracy and precision considerably greater than the accuracy and precision of the operator himself.

At the time it was not a simple matter to measure the height of an airplane above ground with an accuracy greater than that of the stereoscopic heightfinder operator. The human eye can detect stereoscopic differences of the order of a few seconds of arc. The heightfinder magnifies target images and extends the effective distance between the eyes to a number of feet. Thus, when the heightfinder is properly maintained and operated it is far more accurate and precise than any common measuring instrument. Radar gear of sufficient stability and accuracy to give a better measure of height than the heightfinder was just developing at the time of the first Panel studies and was not commonly available until much later.

When it is difficult to measure the accuracy of human performance, it is common to concentrate on the variability. In the case of heightfinder operators this was one solution adopted for the research of Division 7. A plot was made of the successive height readings for any single course of a target. A smooth curve was drawn through the plotted points and the variability of the points around the smooth curve was calculated.

In the use of this method the significance of the neglect of accuracy was, and still is, problematical. It has been routine service practice to neglect constant errors in original firecontrol data. Constant errors can be corrected in combat firing, either during the solution of the firecontrol problem or by spot corrections to successive salvos, but no simple method exists to correct variable errors other than arbitrary smoothing procedures. Variable errors introduce high frequency errors into the computer system, and in many computing systems high frequencies are amplified to a greater extent than low frequencies. Thus some sort of variability score was essential in the measurement of heightfinder operation.

On the other hand constant errors have become more and more undesirable in modern war. Such errors always waste time in firecontrol. When targets move at high speed an anti-aircraft battery may have time for only a few shots before the target has done its damage and passed on. Multiple attacks by many planes at once intensify the need to save time.

Division 7 therefore set about to improve the Army's standard photographic triangulation methods of measuring the "true" height of the target at all times during the practice course. Knowing true height, both the accuracy and variability of the operator and his instrument could be determined.

The photographic triangulation method used by the Army required a phototheodolite at each end of a measured base line some thousands of feet in length. The telescopes of the phototheodolite were tracked on the target of the heightfinder operator. At intervals, synchronized with the height readings of the operator, the angle scales of the telescopes were photographed. This permitted triangulation on the target and the determination of its "true" height.

Since the tracking of the telescopes was not perfectly precise, it was arranged that the photographs of the telescope dials include a view of the target taken through the telescopes of the phototheodolites while the tracking proceeded. These target pictures provided the necessary corrections to the other photographic readings and permitted computation of a satisfactory measure of true height.

The phototheodolite method thus came to furnish a measure of the proficiency of the heightfinder operator which was itself of the highest face validity and of satisfactory accuracy. It remained to determine the consistency of the operators in reading height. For 22 courses run by each of 92 men during 10 to 14 days at the end of school training in the Army, the reliability[2] of the accuracy criterion score proved to be .68. The corresponding figure for the variability score was .82. The sum of the accuracy and variability scores had a reliability of .87. Reliabilities of this order of magnitude were obtained, however, only when instruments were maintained in

2 First half—second half, uncorrected.

first-class order, calibrated properly, and operated by the best available procedures. Under any other conditions the personnel factor was hardly worth measurement since the instrumental errors became much larger than personnel errors.

For the same group of men the accuracy and variability scores were related to one another as shown by a correlation of .80.

The phototheodolite method was slow, difficult to use, and expensive. When applied as a way of determining the proficiency of student heightfinder operators in their final examinations at Camp Davis, some thirty men were required simply to read the phototheodolite records and calculate the scores. For this reason when research on the rangefinder operator began, the project staff operated under a strict Panel directive to find an adequate method of measurement of rangefinder proficiency which could be routinely used in school and refresher training without such a cost in manpower. The project spent considerable time in investigating the actual accuracies of various radar sets, not in the sense of a theoretical limit of accuracy, but in the sense of the accuracy resulting when the radar sets were routinely used, and with routine maintenance and calibration. Shortly before the project ended, training schools began to receive a radar set which was sufficiently accurate to furnish the basis for simple easily-used performance tests of the stereoscopic rangefinder operator.

Checksight Methods

A major problem in all the Panel projects on antiaircraft gunnery and firecontrol was to determine the nature and magnitude of errors in tracking an aircraft target. For any gunnery system a major source of error, assuming that calibration and maintenance were adequate, was error in tracking.

Again however, as in the case of the heightfinder, the human being is a remarkably accurate device when he works with good equipment. When the Panel began its work three methods were in common use for the evaluation of tracking. The first was quantitative and the other two were qualitative. From a gun-camera photograph of the gunner's reticle and field of view, the error in tracking could be measured. From direct observation of a gunner's behavior and the movements of his gun, officers and instructors obtained a qualitative picture of the man's performance. In the Navy this picture was made somewhat more valid by use of a third procedure, involving a device known as the checksight. The checksight was a supplementary telescope aligned with or viewing through the gunner's own sight. With the checksight an instructor could observe the nature of error in tracking.

None of the three methods of judging performance was satisfactory for psychological purposes. The gun camera gave continuous difficulty in the mere matter of adequate photography alone. In a study of gun camera films at a B–29 operational training center R. R. Sears found that less than half of the films were of a quality such that the film could be scored. To score the remainder required six man-hours for the film exposed in each six-minute period of actual practice in tracking and ranging with the B–29 gunsight. By an ingenious adaptation of the checksight scoring methods to be described below, Sears cut this scoring time to one third of its previous value with little loss of scoring reliability. Even so the scores for a large number of gunners could be obtained only after an immense amount of labor and only some time after any experimental or training session.

The delay in providing gunners with knowledge of their performance made the gun-camera method of doubtful value in training gunners. Sears found no evidence of improvement in gunner performance over a series of aerial missions during the operational training of B–29 gunners. A negligible improvement (although one which was statistically significant at the 5 per cent level) was found in gun-camera training on mock-up devices.

Qualitative observation of gunner behavior and gun movement was also proved unsatisfactory as a basis of judging gunner performance. The accuracy and reliability of these judgments were studied for three qualified antiaircraft officers by J. R. Rapparlie. The officers independently ranked the performance of each of thirty enlisted gunners on four practice courses. For each gunner, two courses were run in immediate succession early in the experiment and two in immediate succession late in the experiment. The officers attempted to pick the better of the two courses in any successive pair. Simultaneously, a more objective measure of the gunners' performance was obtained by the use of a checksight-scoring method (described below).

The three officers did no better than chance in ranking the men. And in picking the better of two successive courses the officers were wrong as often as they were right; in various comparisons they were wrong in from 43.8 to 54.2 per cent of the cases. Thus the officers' judgments were of no value.

The adaptation of the checksight to score tracking performance was therefore undertaken. In the project on the antiaircraft battery W. C. Biel and his colleagues developed two checksights—one used a standard Army telescope with a special scoring reticle; the second was made of plexiglas, again with a special reticle, for rough and ready use in the field.

The scoring reticle defined tolerance limits within which the gunner had to keep the target if his tracking was to be defined as "satisfactory" or "on target." With a simple stop-clock system an observer at the checksight determined the per cent of total time for a course that the gunner kept the target within the tolerance limits. It may be noted that this is an accuracy type of score.

Percent-time-on-target scores from a checksight were sufficiently accurate and reliable for most psychological purposes, always provided of course that the tolerance limits were reasonably well adjusted to the group of gunners under study. A. S. Householder and W. E. Kappauf pointed out and applied the principle that the tolerance limits should be set in terms of gunner performance rather than in terms of absolute standards if the checksight score was to be useful in training or in psychological experimentation. The most generally useful set of tolerance limits is the one which determines that the average gunner of any group under consideration receive a score of 50 per cent. These limits provide for the most reliable discrimination between all the gunners of the group. Following the work of Householder and Kappauf the Bureau of Ordnance modified the scoring systems of several of its synthetic gunnery trainers to provide for this type of scoring.[8]

When the tolerance limits were properly set, it was shown that percent-time-on-target scores agreed closely with camera records of tracking accuracy. The average error of a percent-time-on-target score from a checksight for a single short course turned out to be only 8 per cent, and when five such courses were averaged together the error was only 4 per cent. It was further shown that different enlisted men, serving as checksight observers, agreed with one another and with camera records in judging the performance of gunners, giving correlations of .79 to .92.

The work on checksights resulted in the adoption by the Army of the plexiglas checksight and installation by the Navy of built-in checksights in its new directors. Directives for the routine use of checksights in accordance with Panel recommendations were issued to all field units.

Near the end of the war an instrument to replace the checksight and furnish scores and coaching signals during the tracking performance was developed by the Bureau of Ordnance. The device was turned over to the Panel for evaluation and further development. In essence the device consisted of a searchlight to be mounted on a target plane and a telescope, phototube, amplifier system, and relay to be mounted on and

---

[8] William E. Kappauf, *Notes on the Design of Phototube Scoring Devices for Tracking Trainers.* Informal Memorandum No. 9, Project N–111. May 23, 1945. Brown University. Washington, D.C., Applied Psychology Panel, NDRC.

aligned with the gunsight. When the gunner was on target, the searchlight was supposed to activate a clock and buzzer and thus the system might replace the checksight operator.

The instrumental reliability of the system was determined by J. D. Coakley. Since the reliability was inadequate, Coakley improved the device with the help of Brian O'Brien of Division 16, NDRC, the Airborne Instruments Laboratories of Columbia University, and the General Electric Company. The result was a device of great potential value for training in tracking. If the target searchlight was pointed at the gunsight within ±10 degrees, scoring reliability was entirely adequate for airborne gunnery systems and perhaps for antiaircraft systems up to ranges of 4,000 to 8,000 yards. The reliability closely approached the theoretical limit imposed by the change of effective light intensity with change of target range. Since this degree of reliability is adequate for many purposes and since the device provides automatic signals of incorrect performance, as well as a score, it should be very helpful in future gunnery training. At the end of the war the device was turned over to the Naval Research Laboratory for validation as a trainer.

Performance on Field Artillery Equipment

In connection with research on errors in the use of field artillery gunsights it became clear that a complete recording system for the detection of personnel errors was desirable. L. C. Mead, L. V. Searle, and K. S. Wagoner, working with the Armored Medical Research Laboratory, undertook the development of remote recording methods for field artillery.[4]

Remote recording was desirable in order not to interfere with normal operating conditions by recording during the experiment. Hence repeating systems were developed for at-

[4] Richard N. Berry *et al.*, *A System of Automatic Devices for the Detection and Recording of Errors in a 105 MM Howitzer Battery*. OSRD Report 5313. July 6, 1945. Tufts College. Washington, D.C., Applied Psychology Panel, NDRC.

tachment to the four guns of a field artillery battery and their instruments. These transmitted the position of the gun tubes and the setting of the instruments to a recording center located in a truck at a distance. Spotting methods were developed to check the accuracies of fire itself and of the data from the forward observation post. All communications were recorded. The accuracy of recording was adequate since the errors of recording were smaller than the errors in field artillery operations of interest to the Panel.

This recording system suffered from one defect, the correction of which would have required a change of normal field artillery practices for the duration of any experiment. The recoil of field artillery guns changes their position after each round fired. No method was devised to give a remote record of changes in the position of the gun carriage. Hence in an experiment the guns would have had to be more firmly emplaced than is customary in field artillery practice. The fact that the deficiency in recording had to be controlled by a temporary change in normal procedures suggests a need for a permanent change in the normal procedures or in the guns themselves. Movement of the gun carriages after each shot must be a source of loss of precision (or of loss of time) in firing.

The recording system for field artillery was completed in the summer of 1945 at the termination of the work of the project. The system was offered to Army Ground Forces for further and more complete experimentation on field artillery errors. Since the Army was not interested the Panel sold the materials for salvage.

Recording the Performance of B–29 Gunners

The most ambitious effort of the Panel to develop a method for systematic rapid measurement of all significant elements in gunner performance on firecontrol equipment occurred in research on B–29 gunnery. C. H. Wedell was responsible, carrying on the development with the help of the project staff and of the psychologists of the Research Division at Laredo Army Air Field. Wedell set up a complete ground mock-up of the B–29 gunnery situation with remote recording of various aspects of gunner error.

One of the B–29 gun stations was duplicated. The gunner operated his sight against a target, which at first was a projected image[5] and later was a special model airplane. The target was remotely controlled by cam drives so that it could appear to make realistic "attacks" on the gunner or, if desired, so that the various components of target motion could be systematically varied in independence of one another. Gunner performance was recorded remotely in several ways:

A. The gunner was considered independently of his effects on the firecontrol equipment in the following records and scores;
   1. Error records on a polygraph which showed the difference between true target position and gunsight position for each of the components (azimuth, elevation, and range) separately. The moments of triggering could also be recorded.
   2. Time-on-target scores for tracking and ranging.
   3. Integrated average error scores for tracking and ranging.
B. The gunner was considered in terms of his effects on the firecontrol system in the following records;
   1. The difference between a computer which was driven by perfect tracking and ranging (target input) and a computer which was balanced statically and dynamically with the first computer but which was driven by gunner tracking and ranging.
   2. Variability of computed lead.

All scores were calculated instantaneously and automatically. This not only permitted convenient rapid experimentation on

---

[5] This was not a motion picture image. Motion picture targets were rejected for experimental purposes because of the inherent inaccuracies of motion picture projection relative to gunner inaccuracies.

a large scale but it allowed adequate knowledge-of-results training within the experimental situation. It was Panel experience that service training was not sufficiently standard to be used as a basis for equating groups of gunners for experimental purposes. It was necessary that the Panel itself undertake to train experimental subjects.

The ground mock-up developed by the project is shown schematically in Figure 12 for the case of the projected-image target. At the extreme right is a screen (A) on which the target image was projected. A gunsight was mounted in the replica of the blister station of the B–29 airplane at (B). The target source was at (C) where special mechanisms eliminated the parallax between the target projector and the gunsight. The cam drives at (D) moved the target projector through a remote control system. At (E) the data on true target position was compared with data from the gunsight to give the various measures of gunner performance named above.

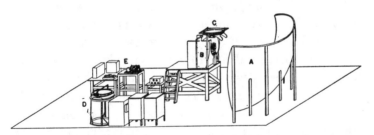

Figure 12. The ground mock-up for the study of aerial gunners and gunsights. See text for explanation.

This ground mock-up was developed in order to experiment systematically on B–29 and other airplane gunnery equipment. The project believed that systematic airborne experimentation on personnel factors affecting the design of gunnery equipment would be very difficult, since the number of uncontrolled variables which affect airborne experimentation is large. The coordination of several airplanes in an experimental military situation is difficult, and it was common experience during the war to obtain reasonable data from full-scale airborne trials only if a great many more trials were scheduled than were actually needed for final analysis. Airplanes broke down, the weather intervened, coordination aloft was not always achieved. In order to obtain the number of observations required for reliable measurement of a reasonable sample of personnel, an enormous number of flights had to be scheduled. The project concluded that full-scale airborne trials should be used only as a rough empirical check on conclusions more exactly established in the somewhat artificial conditions of a laboratory mock-up.

Nevertheless the solution of realistic ground recording of gunner performance provided a pilot model of an airborne test and training device. Most of the variables disturbing experimentation and training aloft come from outside the gunner's own plane. A target carried within the gunner's own plane would eliminate many of the variables. A remotely controlled target and recording system for airborne use was developed by D. G. Ellson from the project's ground mock-up.

The target model of the ground mock-up was so mounted in the side blister of the B–29 that the regular gunsight could be used in a normal way to track and range on the model. The apparent motion of the model was remotely controlled. The ground recording and scoring devices described above were adapted to operate under airborne conditions. This device was never actually tried out aloft by the project since it was not fully completed at the end of the war. Nevertheless ground trials indicated its probable success in the air. Like the ground mock-up it was turned over to Army Air Forces when the project ended.

Solution of the airborne recording problem provided an ingenious polygraph which should be of general value in psychological research. The polygraph[6] has these interesting

6 K. U. Smith, *A Graphical Recorder for Synchronous Linear Registration of*

characteristics in recording motions of frequencies below about 5 cycles per second:

1. Linear registration of four separate motions is provided. Additional on-off signals can be added as needed.
2. The entire chart width of 12 inches is available for all four motions; yet each one of the four records has exact point-by-point simultaneity with the other three.
3. Any one of the four channels may be expanded indefinitely in sensitivity regardless of the size of the motion to be recorded.
4. The device is highly resistant to shock and vibration.

From these test devices came numerous others, among them a synthetic ground trainer. The ground mock-up was simplified by the project and the Research Division at Laredo to provide the pilot model of a realistic ground trainer with adequate immediate scoring of gunner performance. The pilot model was used in experiments on equipment design (see below) and the Army Air Forces prepared from it a preproduction model of a finished synthetic trainer. Studies of the device as a training instrument were to begin shortly after the date at which the project terminated.

EXPERIMENTAL RESULTS

The methods of measurement described above were applied to a host of personnel problems in the field of firecontrol. Selected experimental studies will be described in the following sections of this chapter. The choice of studies to be described here has been determined either by the actual importance of results in World War II or by their potential importance for further research in military psychology. Samples from each of the fields of aptitude, training, and equipment will be included.

Several Mechanical Movements. OSRD Report No. 5937. September 26, 1945. University of Wisconsin. Washington, D.C., Applied Psychology Panel, NDRC. The Phipps and Bird Co. of Richmond, Va., developed the polygraph.

Aptitude

Aptitude tests for firecontrolmen were most extensively studied in Panel research in the case of the operators of stereoscopic heightfinder and rangefinder equipment. As indicated above, the Panel followed up the work of Division 7, NDRC, by validating visual tests for the criteria of accuracy and variability of stereoscopic readings of height and range. The series of studies also provided evidence on the reliabilities and interrelations of a number of visual tests. In general the tests for visual acuity, phoria, and stereoscopic depth perception had rather low reliabilities, and tests purporting to measure the same function often correlated significantly below their reliability.[7] The Panel also determined the validity for the stereoscopic operator of the paper and pencil tests of the U.S. Navy Basic Classification Test Battery.[8]

The paper and pencil tests of the Basic Battery proved to have validity coefficients of roughly .50 for school grades in the academic and shop courses for rangefinder operators. They had no validity for performance in taking ranges.

The visual tests validated by the Panel included standard medical tests for acuity and phoria and a variety of tests of stereoscopic sensitivity. In both the Army and Navy studies the samples were small but the results of the two series of

[7] J. K. Adams, H. M. Fowler, and H. A. Imus. *The Relationship of Visual Acuity to Acuity of Stereoscopic Vision*. OSRD Report 2087. September 15, 1943. Brown University. Washington, D.C., Applied Psychology Panel, NDRC.

J. K. Adams, D. C. Beier, and H. A. Imus, *A Test-Retest Reliability Study of the Bausch and Lomb Ortho-Rater with Naval Personnel*. OSRD Report 3969. August 1, 1944. University of Wisconsin. Washington, D.C., Applied Psychology Panel, NDRC.

H. M. Fowler, H. A. Imus, and F. A. Mote, *Interrelationships Among Seven Tests of Stereoscopic Acuity and the Relationship Between Two Tests of Visual Acuity and Two Tests of Phorias*. Memorandum No. 12, Height Finder Project. March 24, 1944. Brown University. Washington, D.C., Applied Psychology Panel, NDRC.

[8] D. C. Beier and Florence Gray, *The Selection of Fire Controlmen (O): Rangefinder and Radar Operators*. OSRD Report 4861. March 26, 1945. D. C. Beier *et al.*, *A Follow-up Study of the Efficiency of the Projection Eikonometer Test in Predicting the Performance of Stereoscopic Height Finder Observers*. OSRD Report 4352. November 21, 1944. University of Wisconsin. Washington, D.C., Applied Psychology Panel, NDRC.

studies were in close agreement, and the results in each service were consistent for each of several successive school classes.

The results indicated a reasonable degree of validity for a number of stereoscopic tests. The best was the Stereo-Vertical Test of the Projection Eikonometer, a test developed by H. A. Imus.[9] This was a simple test of stereoscopic discrimination for a moving target in which two moving lines were presented, one to each eye. These lines, in fusion, appeared as a single line rotating toward and away from the testee in the vertical dimension about the midpoint of the line. When the line appeared to be vertical the subject pressed a key. Scoring was in terms of the variability of the judgment. The test was individual but could be given to six men at once in a multiple arrangement.

Of lesser validity but nevertheless of a validity adequate for use was the Ortho-Rater stereoscopic test. Several other tests developed by W. F. Dearborn and P. W. Johnston appeared hopeful for possible further research. Combinations of these stereoscopic tests had little greater value than Imus's Stereo-Vertical Test alone.

In Figure 13 is shown the proportion of operators ($N=92$) receiving given scores on the Stereo-Vertical Test who performed satisfactorily on the stereoscopic heightfinder. Satisfactory operation was defined as taking height readings whose accuracy and variability combined would permit solution of the firecontrol problem with an established degree of precision. On this definition, two-thirds of all the students turned out to be satisfactory. The curve shows that performance on the test is related to later performance as a heightfinder operator. The data correspond to a validity coefficient of .50.[10]

[9] H. A. Imus, *Manual for the Adjustment and Operation of the Projection Eikonometer.* OSRD Report 1340. March 29, 1943. Brown University. *The Relationship Between Test Scores Obtained on the Single and Multiple Projection Eikonometers.* OSRD Report 1789. August 5, 1943. Brown University. *Manual for the Installation and Adjustment of the Multiple Projection Eikonometer.* OSRD Report 4233. October 10, 1944. University of Wisconsin. Washington, D.C., Applied Psychology Panel, NDRC.

[10] Product-moment correlation; $N=92$. Since the test and criterion scores

Results essentially similar to those shown in Figure 13 were obtained for the rangefinder operator. As a result of these studies the Panel's recommendations for the selection of Army heightfinder operators and Navy rangefinder operators were adopted and used by the two services.[11]

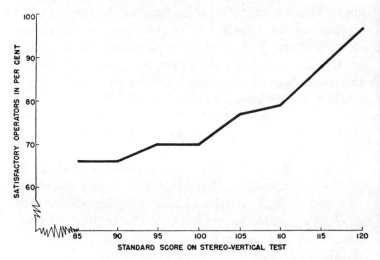

Figure 13. The relationship between scores on the Stereo-Vertical Test and performance on the stereoscopic heightfinder. The proportion of men who are satisfactory on the heightfinder increases with score on the test. $N=92$.

Training Firecontrolmen

Knowledge of Results. In a series of experiments in several of its projects the Panel demonstrated the value to training of informing the student of his progress. In many cases, the hasty organization of wartime training and the difficulty of measuring progress prevented the application of this well-known educational principle. The Panel demonstrations of the

were not distributed normally in raw-score form, the scores on both were transformed to logarithmic scores for correlational purposes. The group was representative of the normal Army distribution on visual acuity.

11 H. A. Imus, *Manual for Use in the Selection of Fire Controlmen (O),* (*Stereoscopic Rangefinder Operators*). OSRD Report 4050. August 22, 1944. University of Wisconsin. Washington, D.C., Applied Psychology Panel, NDRC.

validity of the principle were generally but not always successful. In most studies results were striking enough to secure immediate changes in military practices, but in some instances the differences between expected and obtained results suggest the need for continuing research if military training in the future is to be most effective. War conditions prevented an experimental analysis of the reasons for the occasional failures to prove the principle.

One successful demonstration of the efficacy of knowledge of results in training was conducted by J. H. Rapparlie. Rapparlie organized three groups of ten student gunners each to learn to track with a computing gunsight for the 40-mm. gun. On this gunsight two trackers are required; one tracks in azimuth, the other in elevation. One of the three groups of subjects in the experiment was trained by an officer using the standard Army training method of verbal coaching before, during, and after practice. The second group used a second method, named Co-Tracker Guidance, developed for use where checksights were not available. In Co-Tracker Guidance one tracker coached the other tracker on alternate courses. For the third group of gunners the telescopic checksight operator sounded a buzzer when the gunner was off the target. The buzzer could be heard by many of the gunner's associates and aroused considerable spontaneous interest and competition. The performance of all gunners was scored by the checksight method but only the checksight-trained group of gunners were given their scores at the end of each course.

The performance of the three groups throughout the practice period is illustrated in Figure 14. For each trial of each group the average per cent time off target is shown. It is evident that the three groups were equal in tracking ability at the start of the experiment but that the checksight group learned more rapidly and reached a higher performance level than either of the other groups. Standard Army training was the poorest method of the three. Knowledge of results in the form of accurate immediate information gave the best training in tracking. Further experiments indicated that the advantages of the checksight method of training were retained for at least four weeks after the last practice session.

Team Training. Of particular interest, since it represented the only Panel investigation of methods for training groups of men to operate as teams, was a scoring device indicating

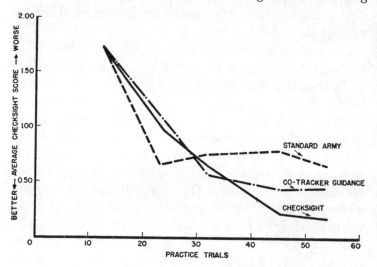

Figure 14. The accuracy of antiaircraft tracking as affected by knowledge of results in training. The Co-Tracker Guidance and the Checksight-trained groups received two forms of knowledge-of-results training while the Standard-Army-trained group received no knowledge of results. $N=10$ for each group.

solution time on the Navy Gun Director, Mark 37. In this director system teams of four, five, or more men must work together at high speed if the enemy plane is to be shot down. If their work is well done the firecontrol problem of predicting future target position is solved rapidly. By a combination of an automatic timing device and a checksight D. G. Ellson developed a sensible measure of the speed of solution. The device could be built into the complex director system without interference with routine operations. When the measure was introduced in a Navy school, rather dull practice sessions were replaced by lively cross-talk and some good-natured betting between director teams occurred.

This type of team-scoring device is basic to effective team training and to future psychological research on teamwork and, perhaps, on leadership. Psychological studies of teamwork in the military field are exceedingly few in number. The paucity of research may well have resulted from the lack of adequate experimental methods for dealing with groups in the military situation. Ellson's device shows that adequate methods can be developed.

Range Estimation. In the cases of many of the lighter weapons of the machine gun and intermediate types it was necessary for one or more gunners to estimate the range of targets in order to determine the moment at which to open fire. In other cases it was necessary to enter an estimated range into computing systems.

It was generally believed that gunners tended to open fire too soon in combat. This tendency would be dangerous since it would exhaust ammunition without effect on the enemy and require cessation of fire at a critical moment while a new magazine of bullets was placed in the gun.

Two methods of range estimation were in common use in the Navy. The first was direct judgment presumably based on the visual angle subtended by the target in relation to its true size. The second was the comparison of apparent size of the target with the apparent size of some fraction of a dimension of the gunsight reticle. For operational purposes it was desirable to know the accuracy of each method.

Several Panel projects investigated the subject. For a criterion measure target range as determined by radar was satisfactory. Gunners estimated the range of an incoming target plane at one or more points in its course, either by guessing the distance at stated moments or by indicating the moment when the target reached a given distance.

In training the men, a procedure similar to that just described was used except that knowledge of results was given.

Men were commonly trained on both real and synthetic targets.

For short opening ranges and for unaided range estimation D. D. Wickens found that, like combat gunners, untrained men showed a strong tendency to make the error of under-estimating opening range.[12] This experimental result, obtained when the gunners were in no danger of enemy fire, indicated that any similar combat tendency could not be due alone to a desire to hit the enemy before he could open fire. In part at least the tendency to underestimate target distance at short ranges must be a reflection of the characteristics of human perception.

The tendency to underestimate opening range was not ex-hibited for targets at medium ranges, according to a study by M. H. Horowitz. At such ranges the average error of range estimation is roughly proportional to range. The error is not very different in size from the error of estimating the range of ground targets. (For ground targets familiar objects are pre-sumably available to furnish secondary cues which might be expected to assist in range estimation.)

In studying the training of men in range estimation, Wickens found that a commonly used synthetic trainer for unaided range estimation permitted the men to learn by secon-dary cues, and as a result had no validity. When the device was modified for training in use of a reticle as an aid to range estimation, it was of value in reducing errors on real targets.

In additional studies it was shown that (1) knowledge-of-results training against real targets decreased the error of range estimation by a considerable degree; (2) range esti-

[12] Morris S. Viteles *et al.*, *An Investigation of the Range Estimation Trainer, Device 5C-4, as a Method of Teaching Range Estimation.* OSRD Report 4263. October 18, 1944. University of Pennsylvania. Washington, D.C., Applied Psychology Panel, NDRC.

M. H. Rogers *et al.*, *Evaluation of Methods of Training in Estimating a Fixed Opening Range.* OSRD Report 5765. September 19, 1945. University of Pennsylvania. Washington, D.C., Applied Psychology Panel, NDRC.

H. A. Voss and D. D. Wickens, *A Comparison of Free and Stadiametric Es-timation of Opening Range.* OSRD Report 6114. October 16, 1945. University of Pennsylvania. Washington, D.C., Applied Psychology Panel, NDRC.

mation with the aid of a circular reticle was only slightly superior to unaided range estimation when the apparent size of the reticle was three times the apparent size of the target at opening range; and (3) refresher training was needed since variability in range estimation increased when training was discontinued. As a result the Panel developed a manual for training in range estimation for use in the Fleet.[13]

The Design of Synthetic Trainers. The Panel's extensive experience with synthetic training aids suggested to the Bureau of Ordnance the need for a general formulation of psychological principles in the design, development, and use of synthetic devices. During World War II such devices were often developed by engineers and put to use without due consideration of personnel needs. In one case an experimental study showed that personnel actually decreased in skill as they practiced on a synthetic device. In a number of cases such devices were found to be harmless but ineffective. In a great number of instances it was found that a fundamentally good device could be improved. In response to the request from the Bureau of Ordnance each of the Panel projects concerned with synthetic trainers wrote up its suggestions for their use, design, and improvement. These suggestions were incorporated by Dael Wolfle in a Panel report.[14] Wolfle's report gave a thorough discussion of the characteristics of good trainers, their use, their advantages and disadvantages in comparison with training on real equipment, the psychological problems connected with scoring performance, and concluded with two check lists covering the psychological steps in designing and evaluating synthetic trainers.

[13] D. D. Wickens, J. H. Gorsuch, and M. S. Viteles, *Memorandum on Manual for Instruction on Mirror Range Estimation Trainer, Device 5C-4 (Equipped with Mark 14 Sight Reticle)*. OSRD Report 5925. September 25, 1945. University of Pennsylvania. Washington, D.C., Applied Psychology Panel, NDRC.

[14] Dael Wolfle, *The Use and Design of Synthetic Trainers for Military Training*. OSRD Report No. 5246. July 6, 1945. Washington, D.C., The Applied Psychology Panel, NDRC. Wolfle's report is not available to the general public at the time of this writing.

The Generality of Tracking Training. A fundamental question in all gunnery training is whether training in one situation is useful in meeting new situations. In the case of tracking it is of basic importance to know whether training on one or a few courses is helpful in tracking other courses, and whether training in tracking with one set of controls or one body posture transfers to other controls and other postures. It is generally assumed in the services that positive transfer of the effects of training in tracking occurs. Thus men trained in tracking on one gunnery device are frequently assigned to duty as trackers on another device. Synthetic trainers usually provide a rather limited assortment of courses for purposes of practice, yet the general psychological literature on training suggests that training in one situation frequently fails to be of value in other situations and is sometimes actually harmful. More precise information on the nature of transfer of training in tracking situations is clearly needed.

It was possible for the Panel to make only two preliminary studies of this fundamental problem. The results clearly indicate that the problem is a serious one requiring detailed investigation if service training in tracking is to be efficient.

One experiment was made possible by the Tufts Tracking Trainer.[15] This trainer proved to give as adequate training in tracking on the gun director which it simulated as training on the director itself. Since the trainer permitted ready and valid measurement of tracking performance in this one situation it could be used to study the generality of tracking training in other situations.

Men's performance on the Tufts Tracking Trainer was tested after the men had been trained on other types of equipment. Their performance on the Trainer was compared with that of untrained men and of men trained on the Trainer itself. Groups of radar trackers, heightfinder trackers, and

[15] Bradford B. Hudson and Lloyd V. Searle, *Description of the Tufts Tracking Trainer*. OSRD Report 3286. February 5, 1944. Tufts College. Washington, D.C., Applied Psychology Panel, NDRC.

trackers trained on other directors were thus compared. No group trained on other equipment did any better than untrained men on the Tufts Trainer. This result confirms the suspicion based on general psychological principles that much service training in tracking is too limited to be of general value.

Comparable results were obtained in studies of the transfer of training from one kind of radar tracking to a second kind of radar tracking.[16]

It may be of interest to psychologists to point out that this particular service situation affords a major opportunity for the controlled study of stimulus and response variables and training methods in relation to fundamental psychological problems—the problems of transfer of training and the generality of ability. It seems probable that the military tracking situation is peculiarly suited to studies of this problem, which should be of considerable direct practical interest to the services. It should be a relatively simple matter to obtain large numbers of personnel of known aptitudes and inexperienced in tracking tasks to serve as experimental subjects, and methods for the automatic measurement and registration of performance in tracking are now available. Systematic study of the problem should yield important results—important not only in a practical sense but important also to a basic theoretical problem.

The Design and Operation of Firecontrol Equipment

Interpupillary Distance and the Heightfinder. In research by Division 7 it was demonstrated that large errors in ranging were produced by very small errors in setting the eyepieces of the heightfinder at the interpupillary distance of the operator. The effect was due to parallax between target image and reti-

---

[16] Irving H. Anderson *et al., A Study of the SCR–584 Basic Trainer as a Training Device for Learning Range Tracking.* OSRD Report 3344. February 10, 1944. Yerkes Laboratories. Washington, D.C., Applied Psychology Panel, NDRC.

cle. When the eyepieces were wrongly set, the observer's pupil did not include the entire exit pupil of the instrument. In consequence the apparent range of the target varied in relation to the reticle as the operator moved his eye from right to left across the exit pupil. The size of the error varied with the range of the target.

Previous to this finding the operator of the stereoscopic heightfinder or rangefinder had measured his own interpupillary distance with a hand-held interpupillometer and set the result into the interpupillary scale of the instrument. The effects of erroneous interpupillary settings were not at all apparent to the operator. In consequence he was often careless in measuring his interpupillary distance and setting it into the heightfinder.

Four steps were taken to correct the situation by W. E. Kappauf and his colleagues of the Panel projects on heightfinder and rangefinder operators. (1) Available interpupillometers were investigated to determine their accuracy. None supplied a measure of sufficient accuracy to reduce errors in height or range readings to a negligible size; therefore an interpupillometer of adequate accuracy was built.[17] (2) The accuracy of the interpupillary setting mechanism on the heightfinder was investigated. The mechanism contained so much play that even a careful setting by the operator allowed significant error. (3) A new method, independent of the heightfinder scale, was developed to permit accurate setting. (4) Operators were indoctrinated in the necessity to make this setting accurately. A demonstration was developed for serv-

[17] W. E. Kappauf and W. J. Brogden, *An Instrument for the Measurement of Interpupillary Distance.* OSRD Report 1372. April 6, 1943. Brown University. Washington, D.C., Applied Psychology Panel, NDRC.

J. K. Adams, D. C. Beier, and H. A. Imus, *The Reliability and Precision of the NDRC and Bausch and Lomb Interpupillometers.* OSRD Report 3475. March 29, 1944. University of Wisconsin. Washington, D.C., Applied Psychology Panel, NDRC.

D. G. Ellson and H. A. Imus, *Memorandum on the Construction and Calibration of the NDRC Interpupillometer.* OSRD Report 3797. June 19, 1944. University of Wisconsin. Washington, D.C., Applied Psychology Panel, NDRC.

ice schools showing the effects of erroneous interpupillary settings on accuracy in reading height. Standard procedures for making the setting were introduced into training manuals and other literature.

The interpupillometer and method of setting the height-finder to the operator's interpupillary distance were adopted by the Army and Navy, and the men of both services were given special training in their use.

Gun Battery Alignment. In the operation of antiaircraft systems a frequent and critical problem arises in the alignment of various units of the system. The guns and gun director, for instance, are often mounted independently of one another. Unless the alignment of each unit with the other is satisfactory, it is obvious that the system as a whole will be inaccurate. The procedures by which alignment is accomplished become quite complex if, as is common, more than one gun is tied in with a gun director or more than one gun director with a gun. F. V. Taylor was charged with the preparation of standard procedures in this as in other matters concerning the operation of certain of the Navy's lead-computing gyroscopic sights and directors.

As a basis for a general discussion of alignment, an investigation was made of the problem of achieving sufficiently precise alignment in a radar system. Each of three experienced Navy men independently judged when alignment was correct. Over a large number of trials it was observed that the men disagreed with one another in individual determinations by as much as three to five times the tolerable misalignment. Further analysis of the data showed that to have 95 per cent confidence of satisfactory alignment, as many as 30 independent observations of alignment were required. This number of observations would have required many hours of work.

Firecontrol personnel generally fail to recognize the operator problem in alignment and in similar calibration procedures. Alignment is conceived as a simple process of adjustment and readjustment and it is assumed that each readjustment improves the alignment until it is finally right. Psychological data on errors of observation, however, suggest that there is just as great a chance of error in each observation as in the one before. If this is generally true for alignment, and there is reason to believe that it is, then each adjustment should be made independently of the others and the final setting should be the average of a number of such independent settings.

In connection with this and other calibration problems it was a common Panel experience to observe that firecontrol personnel not only do not understand the necessity for repeated independent measurements, but they even tend to neglect repetition when it is a part of standard operating doctrine. The tendency is strengthened by equipment which can be calibrated only by a tedious and difficult process, and it is sometimes true that time is not available in combat or between battles to calibrate firecontrol equipment properly. Much could be gained by greater attention to personnel needs and capacities in designing equipment for simple accurate calibrating procedures.

Field Artillery Sights. In considering the sources of error in the fire of field artillery the Panel concentrated its major efforts on the recording system described above. Nevertheless the project staff, in cooperation with the Armored Medical Research Laboratory, also approached the problem by a direct count of errors in firing or in "dry" (nonfiring) practice.

In counts of field artillery errors, it was observed that errors tended to be large in numbers at the battery and in communications. Even when errors of less than 3 mils were disregarded, there were still too many errors. The character and source of these larger errors were revealing. Data from several studies conducted by J. P. Nafe and his assistants may be summarized as follows. Of 100 errors (larger than 3 mils) made by the executive officer in aligning a battery of four guns, roughly one-third were 100 mil errors. Of 100 errors of gunners, roughly one-third were in range and fuze, and in adjustment for the relative height of guns and targets. The remaining two-thirds of the errors at the guns were in deflection and elevation. The deflection and elevation errors broke down into the following categories:

> 40 per cent apparently due to confusion and faulty communications.
>
> 60 per cent apparently due to errors in the use of gunsight scales.

Errors in the use of scales subdivided into:

> 4 per cent 100 mil errors.
>
> 68 per cent due to improper rounding of numbers, transpositions of numbers, reading scales in wrong direction, etc.
>
> 28 per cent with no obvious explanation.

These errors in field artillery fire were "human errors." They were made by the officers and men. Although systematic studies of human characteristics predisposing to the making of errors were not carried out, several small samples demonstrated that experienced men and men of high scores on the Army General Classification Test tended to make fewer errors than other men. Thus the errors were truly human since they could be controlled by variation of the human element. Nevertheless they varied with the nature of the equipment in use.

The standard field artillery deflection gunsight for the guns studied by the Panel is shown in Plate 5. The gunner looked through the eyepiece at the bottom of the figure and aligned its reticle with an aiming point. The setting of the gunsight was shown on two scales. On the coarse scale thousands and hundreds of mils were read (02 in the illustration). The fine scale was the outer one of the pair of scales at the left of the sight; on it tens and units were read (37 in the illustration). The reading shown is 0237 mils. The inner one of the two scales at the left was a supplementary scale known as the gunner's aid. On it the gunner could set a small deflection shift and then turn the sight to zero this setting. In this way mental arithmetic could be avoided for small numbers, but there was reason to believe that gunners occasionally read the gunner's aid when they intended to read the fine scale.

The gunner's job required that he read the scales, set the sight to new scale readings on order of the executive officer and make corrections to settings as ordered. Since the four individual guns of a battery required different base settings, the executive officer called out the amounts of corrections to settings rather than the new settings themselves. In consequence the gunner had to perform mental arithmetic on the scale reading as a base number, adding and subtracting one-, two-, or three-place numbers to the reading shown. Frequently a series of corrections was required in one successive set of operations so that two or even three successive corrections of the same or opposite sense might be required in a single series of orders. And different commands might mean the same shift in deflection; e.g. a command to the battery, "On Gun 1, Open 4" meant the same thing to Gun 2 as the command, "On Gun 4, Close 2."

Some sources of gunner confusion and error can readily be seen merely by considering the characteristics of the sight shown in Plate 5. The division of the course and fine scales complicates the task of the gunner. Reading tens and units to the left of thousands and hundreds violates life-time reading habits. In addition the various scales in common field artillery use did not all progress in the same direction. Thus it was easy, in the confusion of firing to read the setting shown in Plate 5 as 0243, as 0337 or 0343, or even as 0537 or 0543. Hundred mil errors were further encouraged by occasional misalignments of the coarse scale. In such cases mental allowance had to be made for the misalignment. So several possible sources of the 100 mil error were obvious enough; the characteristics and locations of the coarse and fine scales of the gunsight produced the error. Certain proof that the scales predisposed to gunner error was sought by obtaining or con-

COARSE SCALE –
REQUIRES INTERPOLATION
BETWEEN UNNUMBERED MARKS,
MOVES CONTINUOUSLY WITH
FINE SCALE, HAS WIDE
ENAMELLED, SCALE MARKS
(POORLY DEFINED)

FINE SCALE –
LOCATED TO LEFT OF
COARSE SCALE, THUS NOT
IN THE NORMAL ORDER
FOR READING

PLATE 5. A field artillery gunsight. The correct reading of the scales is 0237 mils.

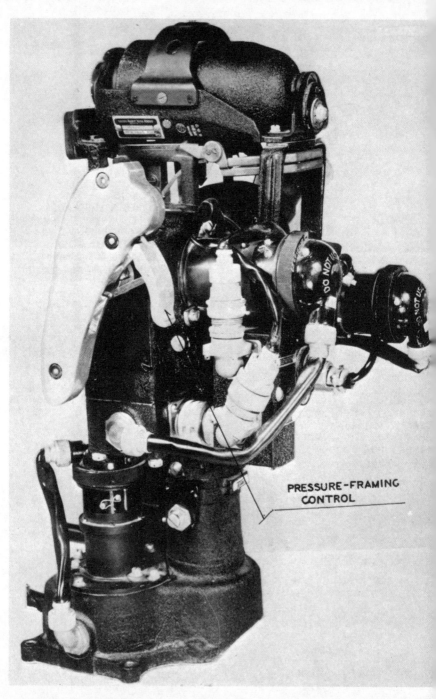

PRESSURE-FRAMING
CONTROL

PLATE 6. Modified right hand control for the B-29 Pedestal Gunsight. Rangin[g]
or framing of the target wing-tips, is accomplished by finger pressure rather than b[y]
movement of the hand as a whole (compare Figure 11, page 155).

structing other artillery sights with different kinds or locations of scales.

A captured German gunsight was studied as one alternative to our gunsight. The scales of the German gunsight were very clean and well-engraved, in contrast to ours. And the German scales, although divided, were more nearly in the customary reading position. In comparative runs, each of 18 American gunners made 100 settings on the German sight and 100 settings on our M–12 sight. They made only 60 per cent as many errors of all kinds in setting the German gunsight as they made on our own familiar American gunsight. And 79 American gunners made only 17 per cent as many errors of 100 mils or more on the German sight as they made on our sight in approximately 1,250 readings on each of the two sights.

For further experimental purposes the Armored Medical Research Laboratory modified the American gunsight by expanding its scales in size and by moving the fine scale to a position more nearly normal for American reading habits. On this modification only 6 per cent as many errors of 100 mils or more were made in reading as on the standard gunsight.

The most hopeful method of reducing errors in reading and setting field artillery gunsights was generally agreed to be to consolidate the divided scales into a single scale. Following a suggestion of W. S. Hunter, L. A. Riggs of Division 7, NDRC, developed an odometer-type sight with a single scale. A pilot model of this sight was constructed for trial by the Army Ordnance Department; unfortunately the Ordnance Department made certain design changes from Riggs's model which defeated part of the purpose of the original design. On test, this pilot model gave only 25 per cent as many reading errors but 143 per cent as many setting errors as the standard American gunsight. The setting errors were believed to arise from the Ordnance Department's modification of the design. At the end of the work of the project a second pilot model was under construction in strict accord with Riggs's design.

The research on field artillery scales is of practical interest chiefly in terms of the future. Whether or not its potential practical values will be realized rests with the Army Ground Forces. From the psychological point of view its present interest is chiefly in the fact that it demonstrated that errors which may properly be called "human errors," since they vary with aptitude, training, and experience, change in character and number with the design of equipment. In one sense the conclusion is obvious, but in many instances this obvious conclusion has been ignored. It is probably ignored at the present time by the use of such terms as "pilot error" in explaining airplane crashes. One may well raise the question: What proportion of pilot errors would be eliminated by changing the design of the airplane cockpit?

B–29 Gunnery. While developing its test equipment for systematic studies of aviation gunnery, the Panel completed several experimental studies of the B–29 gunsight in relation to the gunner. Two may be described here. K. U. Smith, J. K. Adams, and R. L. Solomon investigated the problems involved in triggering the gunsight and possible methods of correcting the errors introduced by triggering. D. C. Beier and W. C. H. Prentice compared the efficiency of a modified set of handcontrols with the efficiency of the standard handcontrols for the sight.

Triggering is a complicating feature in the simplest gunnery systems, such as those of the pistol, rifle, or shot gun. In a gunsight of the character of that used on the B–29 airplane (see Chapter 6), triggering was expected to be disturbing. Records of gunner behavior proved the point.

The controls of the B–29 gunsight were connected to a moving-paper kymograph. Records of control of the gunsight were taken for 34 gunners of varying degrees of training. These men were first given additional training on an Army motion-picture trainer; during 23 days each man "fired" a total of 368 attacks from B–29 sights. They then "fired" a total of 80 attacks per man over a five-day test period. For the test attacks, the effects of triggering were analyzed.

One analysis of the records was made by dividing each attack into successive periods of .133 seconds each. Each period was then classified as to whether or not it contained a sharp change in the elevation-tracking record and whether or not the trigger was pressed or released. A sharp change was defined as a point at which the direction of the tracking curve changed by 20 degrees or more. For 10 gunnery-school graduates over a total of 27 attacks each, 18 per cent of the intervals containing trigger action also contained a sharp change in the elevation-tracking record. In contrast, only 7 per cent of the periods containing no trigger action showed such sharp changes ($p<.001$). A similar difference, sometimes more and sometimes less in amount but always statistically significant, was obtained on separate analysis of the records of groups of gunnery-school candidates, ex-combat gunners, and ex-combat officers. Evidently triggering on this sight produced jerky tracking.

Other analyses were made, including, for example, ranging accuracy during periods of "triggering" versus periods of "non-triggering." Measured in arbitrary units, the average error in ranging was 46 for all subjects during firing and 47 during non-firing periods. This difference was insignificant. The result is in contrast to the common and apparently reasonable expectation that a gunner will fire when he is on-target and hold his fire when he is off-target.

The rather unexpected result just stated was clarified by further analysis of data on the characteristics of triggering. The duration of the triggering and non-triggering intervals was measured from the kymograph records. Sample triggering durations are shown in Figure 15 for three gunners, each of whom fired four attacks. The four attacks differed from one another, but the same four were fired by all three gunners. Even a casual examination of the figure shows that each gunner had a relatively constant pattern of triggering duration and that the pattern differed from gunner to gunner. Thus some gunners triggered frequently and some infrequently, but each adopted his own peculiar semi-rhythmic pattern of triggering. The rhythm was by no means precise, but statistical analysis of the data showed that the ratio of time spent in firing to time spent in not firing was relatively constant for each gunner and different from gunner to gunner. Since the firing was more or less rhythmical, it could hardly have been related to the accuracy of tracking or ranging but must have been semi-automatic.

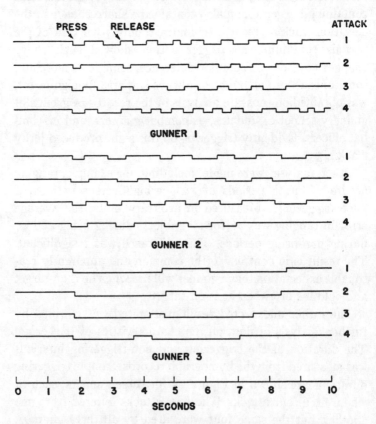

Figure 15. Triggering durations for three B–29 gunners. Each gunner fired against four different attacks. The duration of triggering was typical of the gunner and relatively independent of the attack.

The conclusion of the project was that the task of the B–29 gunner was so complex that its minor components were carried out by the gunner in relative independence of the perceptual situation. The attention of the gunner was so occupied by the tasks of tracking and ranging that he could not make a clear and discriminating choice of the moment at which to fire. In its complexity, B–29 gunnery taxed the limits of human ability.

The practical conclusion from this study was that the operational doctrine should be changed. The gunner should be trained to press the trigger continuously once firing began. If for reasons of materiel it was necessary to fire in bursts, which was doubtful in the B–29 system, a simple lightweight interval-timing device could be installed to permit intermittent fire despite continuous triggering.

The experimental study of triggering, as well as a number of general psychological considerations, indicated the possibility that gunner performance, and particularly smoothness of tracking, might be improved by a modification of the B–29 gunsight controls. The following types of change seemed to be indicated:

1. Reduction of interference between triggering and other movements.
2. Reduction of interference between the movements involved in elevation tracking and ranging.
3. Increased leverage in elevation tracking.

R. L. Solomon designed a new set of handcontrols incorporating these principles. Simultaneously he included certain other minor changes designed to make the new controls fit the hand more comfortably. The controls were installed on a B–29 gunsight for experimental trial. One of the modified controls is shown in Plate 6.

Because the project's test equipment was not yet ready for use, the modified gunsight was tested on the pilot model of the project's Remotely Controlled Trainer, mentioned above. The tests were by no means complete since the target courses were neither realistic nor a systematic sample of various rates, and since only time-on-target scores were available for use in comparisons. Separate time-scores were taken for accuracy in azimuth, elevation, and range tracking. In addition time-on-target for all three criteria simultaneously was recorded.

The gunners were instructed not to use the trigger during the experiment. This instruction was given because the trend in B–29 gunnery, supported by the triggering study just described, was for continuous fire once a gunner had begun to fire.

Thirty enlisted men with no previous gunnery experience served as subjects. They practiced for 350 seconds per day for 23 days. During the first 11 days a buzzer signal was given to the men whenever they were on target. The men practiced on the modified and standard controls on alternate days, half following an ABBA and half a BAAB order.

The results of all gunners are shown in Table 30, which gives the differences between the sight controls and the confidence levels of the differences. Throughout the experiment the modified sight showed a small but statistically significant superiority in azimuth and elevation tracking, in ranging, and in the three components simultaneously. The value of the new controls, expressed as an improvement over standard performance, ranged from 4 to 22 per cent. In independent postwar follow-up studies of the same controls, conducted under P. M. Fitts at the Aeromedical Laboratory, Wright Field, by J. C. Brown, similar results were obtained.

The Panel project designed several other sets of handcontrols for the B–29 gunsight in a preliminary effort to isolate a few of the variables significant for the design of handcontrols. One of these additional sets was sent by Col. E. M. Day, USA, Commanding Officer of Laredo Army Air Field, to the 21st Bomber Command for combat trial and comment. Col. Day's report[18] of the results is as follows:

[18] Letter, Col. E. M. Day to C. W. Bray, September 11, 1945.

TABLE 30. Comparison of modified and standard controls for the B–29 gunsight. The data represent the mean differences and the p values of the differences between the controls in a time-on-target score for azimuth, elevation, range, and all components simultaneously. All differences are in favor of the modified controls. Results are given for the beginning, middle, and end of the training period.

	Azimuth	Elevation	Range	All Components
Difference (Days 1–4)	14.9	3.5	1.8	4.1
p	.0001	.04	.28	.0001
Difference (Days 12–16)	8.1	4.7	2.7	2.7
p	.0001	.0001	.04	.0001
Difference (Days 17–20)	4.8	2.2	4.5	1.9
p	.002	.03	.0007	.009

"Dr. Brogden has asked me to give you what information we received from Guam and the 21st Bomber Command [in] reference [to] the modified sight controls. . . . These controls, carried by our liaison officer to the 21st, were very enthusiastically received. All gunners contacted from a psychological standpoint much preferred these controls. The 21st Bomber Command ordered a full group equipped with the controls for a full-scale service test. The war ended, however, before completion of this test. . . ."

On the termination of work on B–29 gunsights, K. U. Smith prepared a tentative guide of general psychological principles for the use of engineers in designing gunsights. The guide was based chiefly on *a priori* reasoning and general experience with gunnery, and it was by no means presented as a definitive and final work. It was intended for immediate practical use as a compendium of the best available opinion and as a guide to possible future research. Because it has general interest in all problems of equipment design a few items from it are given here:

1. General Factors in Design.
   a. Organization of the equipment in relation to airplane design. Has the equipment been planned with reference to required

provisions for the gunner? Of special importance are adequate space for the gunner to move around, clear vision for the area of fire, adequate arrangement of oxygen supply, and proper seating provisions.

2. Elimination of Operations Requiring Judgment.

   a. Design of the computing system. Check whether the gunner must learn the nature of the computing system in order to operate the sight. Consider carefully the need for selective use of the computer under different conditions before accepting a system which requires it.

   b. Solution of the slewing-on problem. Reduce as far as possible the need for the gunner to know deflection angles, to judge target direction, and to estimate target angles in order to reduce computer error in slewing on the target. Has the sight system a manually-controlled gyro-caging device for eliminating slewing-on error? A caging device is an important requirement for sights with relatively long solution times.

3. Postural Factors in the Arrangement of Equipment.

   a. Position of manual controls. The manual controls should be located directly in front of the gunner, at least six inches below shoulder level and not more than eighteen inches in front of the torso when the gunner is in a sitting posture. Other designs are not considered satisfactory unless properly tested.

   b. Change in arm and shoulder posture with maximum movement of sight controls. Check the maximum movement of the tracking controls in azimuth and elevation in order to determine whether twisted, unnatural postures must be assumed to make these movements. Power control (rate or position) with reduction in magnitude of control movement is a means of achieving a desired design on some director-type sights. Powered seating aids or sighting stations may solve these posture problems.

4. Motor Coordination Factors.

   a. General configuration of controls. Check whether or not switches or push-buttons, requiring intermittent action during tracking, exclusive of a single trigger, are used on the handgrips. Any intermittent reaction imposed on the manual controls can be expected to disturb tracking.

   b. Interaction of controls. Determine whether the control design is such that errors or irregularities of movement with one control cause errors in another control. The B–29 pedestal sight, for example, is so made that irregularities in elevation tracking may cause change in reticle size due to the interaction of the ranging and elevation tracking controls. Similar control interaction occurs on other systems. The general principle to follow is that different but non-antagonistic movements should be used in effecting the primary adjustments of tracking and ranging. A pressure-type ranging control is preferred to several other types because of this principle.

c. Trigger position. If a manual trigger is used it should be an index finger trigger, since movements of this finger can be made without inhibiting the wrist and arm movements involved in tracking. The trigger should be on the far side of the controls and centered in relation to the elevation axis of the handles. Action of the trigger, which will thus have a position of minimum leverage, should cause fewer irregularities in the tracking. Foot triggers are recommended wherever possible.

d. Smoothing and linkage between the gunner's muscular system and the reticle system or guns. There are numerous methods of smoothing tracking operations and tracking data in order to eliminate the effect of tracking variability on firing. Many of these methods of smoothing produce some degree of breakdown of the normal relations existing between the gunner's visual orientation and visual pursuit reactions and the muscular adjustment which he makes to maintain pursuit. Experience and experimental data at present do not permit recommendations as to the best combination of smoothing functions to employ. Experiments should be carried out on several problems in this field, especially those concerned with the optimum ratios of reticle movement with respect to gun- and sight-control movement, and the utility of different aided-laying factors available. Criteria of evaluation of these factors should include both the amount of training necessary to employ these smoothing aids and the absolute level of proficiency possible with them.

5. Reaction Time Factors in the Organization of Sighting Equipment.
   a. Marking of switches. To minimize response times, clear marking and differentiation of switch controls is essential. Check especially the possibility of using light indicators, color differences, or shape differences for critical switches.

6. Fatigue Factors in Sight Design. The degree to which some of the foregoing factors in design are considered will determine the

extent to which operational fatigue of the gunner can be avoided. Especially important in prevention of fatigue are the factors of glare in the design of reticles, proper seating accommodations, space for some change in posture at the sighting station, and the degree to which sustained reactions are eliminated.

7. Training Factors in Equipment Design. A complicated piece of equipment like a computing sight cannot be used without systematic training. The effectiveness of this training will determine the eventual results achieved with the equipment. In original plans it is therefore important to consider factors which will tend to promote effective mass training. Synthetic training devices are needed for every sight so far designed. In the building of such synthetic devices, it is always desirable that there be a means of securing mechanical or electrical data as to reticle position in azimuth, elevation, and range. Means of securing such information might be incorporated in the original design of the sight. A general psychological analysis should be made of the sight in planning stages in order to determine standard operating procedures and basic methods of training.

The research of the firecontrol projects of the Applied Psychology Panel made many practical contributions to service efficiency in World War II. It showed that the effects of the human factor in the man-machine combination can be measured and that the principles of control of the human being have value in the military situation. In the relatively new field of the design of military equipment in terms of human needs and capacities the Panel worked not only for immediate practical results but progressed steadily toward the formulation of generalizations which should assist in future design. There can be little doubt that when such general conclusions are firmly established they will be of major theoretical as well as practical interest.

CHAPTER 8

ACHIEVEMENT AND PROFICIENCY TESTS

THE various research projects of the Applied Psychology Panel undertook to study many problems which appear on the surface to be quite diverse. The prediction of aptitude in a training school for amphibious personnel seems to have little relation to the study of voice communication over aircraft interphones, and neither of these seems closely related to the development of methods for the study of the design of field artillery gunsights. In many fields of such apparent diversity successful research was completed.[1]

As the work of the Panel developed the common thread in the success of the diverse projects became apparent. Consider the nature of the chief research problem which had to be solved before successful research could be completed in each of these specific cases: Navy aptitude tests, training the radio code operator, and the design of B–29 gunsights.

In the development of Navy aptitude tests it was clear from the start that the formulation of the tests themselves would be a relatively simple matter. A wide variety of aptitude tests had already been developed for civilian use. From the technical point of view, some were good in being reliable, homogeneous, objective, etc. Others were less good, but it was a matter chiefly of skillful but nevertheless routine application of well-known principles to improve them.[2] The difficult problem was to determine which of the numerous alternative tests were actually related to military needs. In the case of the Navy aptitude tests the problem was solved by taking the ability to predict school grades as the criterion of a good test.

---

[1] This chapter is based on *STR* II, Chapter 17, by Norman Frederiksen.
[2] The exception to this statement lay in the problem of choosing or constructing tests which should be valid yet independent of one another.

In the case of training the radio code operator there were several possibilities which were fairly obvious. The distribution of practice may serve as an example. Drill in any subject can be too concentrated; men become "stale" if they work too continuously on a single job. But there is a gap between knowledge of fundamental principles and application. Questions arise: Just how should practice be distributed for the specific job when as many well-trained men as possible must go to combat in a minimum time? Since the distribution of practice causes difficulties in scheduling masses of men and loss of time in shifting from one kind of practice to another, is there any practical value at all in distributing practice? Again a criterion measure is needed; sound answers to the questions require the measurement of men's performance during and after training on various schedules.

In the case of the B–29 gunsight the same problem arose again. The possible varieties of gunsight controls are many. Consideration of the general principles of human behavior may cut the number most likely to be successful to a few, but general principles alone can never permit decision between the few in actual application, and general principles occasionally seem to fail altogether if the precise mode of application is not checked by experiment. Thus a criterion of human performance was required for experimentation on B–29 gunners.

The development of criterion measures, then, characterized the psychological approach to military problems. It was common to psychological studies and it distinguished the psychological approach from that of the personnel administrator on the one hand and the engineer on the other hand. The personnel man applied general principles and common sense; evaluation of results was empirical. The engineer set standards for his materiel and measured the materiel to insure that it met the standards; engineering measurements were limited to materiel. The psychologist developed standards for the man in relation to the materiel and measured man's performance to insure that it met the standards.

This common thread in the Panel's research led to the development of measurement of human behavior for purposes more general than research alone. As the specific projects completed their investigations in the various training camps, their experimental criteria were observed to be useful as methods of examining men in training. The research criteria turned out to be suitable in setting graduation standards. Since the criteria were realistic and not subject to the whims of particular instructors, they motivated trainees to learn. The criteria showed the instructor in which areas his instruction was weak and in which it was strong. Over the years 1942 and 1943 it became apparent that training could benefit from the use of research methods of measurement.

The Bureau of Naval Personnel asked the Panel to assist it in the development of standardized objective measures of the performance of Navy trainees and candidates for promotion. Each of the Panel's Navy research projects was asked to adapt its research methods for routine use in the school or field situation. The Panel accepted the request and began a systematic program of development of achievement and proficiency tests in January 1944. Each of the Navy projects made a contribution. In particular the project on Navy aptitude tests concentrated a large share of its energies on the new problem. Norman Frederiksen, C. M. Harsh, D. M. Peterson, N. Fattu, M. D. Bown, and others under the direction of Harold Gulliksen prepared objective examinations for many Navy schools.[3]

Panel performance tests were of many kinds and varieties. At one extreme was the familiar objective question now used quite frequently in school and college to test verbal knowledge and understanding. At the other extreme were measures of

[3] The final report of the project on Navy aptitude tests gives a summary and bibliography of its work, but none of the project's reports on achievement and proficiency tests is available to the public. Herbert S. Conrad, *Summary Report on Research and Development of the Navy's Aptitude Testing Program: Final Report on Contract OEMsr-705 (September 1, 1942–October 31, 1945)*. OSRD Report 6110. October 31, 1945. College Entrance Examination Board. Washington, D.C., Applied Psychology Panel, NDRC.

skill in operation or maintenance. Such tests are known as achievement tests if knowledge and understanding are emphasized. They are known as proficiency tests if skill is emphasized. Either kind of performance tests measures the specific knowledge or skills gained through specific experience or training on a particular job. They differ, therefore, from aptitude tests which test the knowledge or skills which most of our population have had an opportunity to acquire. The techniques which are used to develop and check the internal characteristics of achievement and proficiency tests are essentially the same as those used in the development of aptitude tests (see Chapter 3). The character of the tests developed by the Panel may be illustrated by describing several examples.

It was commonly necessary that trainees know the names and functions of the parts of their equipment. In a paper-and-pencil multiple-choice achievement test, a function might be described in a sentence. Under the sentence the names of various parts were listed. The trainee was required to pick the name of the part performing the function described in the sentence. The same purpose might be served by showing the parts in a picture instead of by naming them. Or the actual part might be set up in a room in association with a list of parts printed on one card and a list of functions printed on a second card. The task was to match the part with its name and function. One or both of the lists might be changed to lists of adjustments, common repairs, or equipment from which the part might have been taken.

The primary problems for the examiner in the original construction of this type of test were to choose items which were important, false answers which did not mislead the better students, and questions whose answers required knowledge and thought rather than memory alone. Statistical analysis of the items and of the test as a whole checked the examiner's success or failure.

In some schools mechanics were trained to make metal

objects to specifications. Instructors varied in the extent to which they held the students to the specifications. In one school different instructors showed little agreement with one another in rating the same student products; r varied from —.11 to .55. The Panel then furnished each instructor with a set of four simple gauges designed to measure the standard product. This application of industrial gauging methods brought the correlations between the same instructors up to .93 and .94. With the gauges the self-correlations for the instructors, when rating products twice at an interval of 10 days, were .97 to .98. The improvement in reliability resulted chiefly from an increased spread of scores when the gauges were used.

Proficiency tests developed by the Panel were also represented by the radio code tests described in Chapter 5, by the checksight and other methods described in Chapter 7, by the assembly and disassembly of a gun, by throwing a searchlight out of adjustment in a specific way and requiring the student to readjust it, or by requiring him to calibrate the depth control mechanism of a torpedo. When the task was standardized and the performance measured or rated on specific details, these proficiency tests were very useful. It was generally found that the introduction of the tests produced improvement of student performance. The reasons were varied. In some cases the men seemed better motivated. In other cases the instructors changed their ways; preparing their students for proficiency tests required that the students practice rather than listen to lectures.

In one amphibious training base, proficiency tests were introduced after their development had been completed at another. It turned out that some subjects, semaphore sending for instance, were taught very effectively, while other subjects were taught quite ineffectively. At graduation no student was found who could read a course by magnetic compass, nor could a student be found who could receive blinker messages at the standard rate in use in the Fleet. At a radio school two-thirds of the students were found to be unable to tune their sets to a specific station and this despite a number of hours of practice in tuning the same radio sets used in the proficiency test. Investigation showed that the instructor had simply assumed that the men knew how to tune their sets. He had ordered them to practice tuning and then turned his back. The students twiddled dials until a station was tuned in by chance. The facts were unknown to the officers in charge until the proficiency tests were introduced. It was easy to correct such situations when the facts were known.

In other ways, too, a change came about. For example at a basic engineering school the students spent four-sevenths of their time in shop with the remainder of their time divided between mathematics, shop theory, and mechanical drawing. Three examinations were given, one in practical shop ability, one in mathematics, and one in mechanical drawing. It turned out that the students differed little from one another on the shop examinations, whereas they differed greatly in mathematics. The net result was that mathematics counted much more heavily than shop in the final grade although much less time was spent on mathematics than on shop. This kind of situation was controlled by the development of achievement and proficiency tests with more adequate internal characteristics.

Possibly the most important effect of achievement and proficiency tests was that they brought about a standardization of the curriculum in many training centers and provided a means of controlling the quality of the output of trained personnel. In a mass Army or Navy it is just as important as it is in a mass-production industry that the product meet specifications. In either case, whether the product be men or machines, the basic method of insuring high quality is to measure the significant characteristics of the product as accurately as possible.

Thus the introduction of standard objective examinations provided a method for a central quality control over widely separated schools. It was no longer necessary to judge the work of various schools merely by qualitative impressions or by "brass hat" inspections. When one had a standardized intelligibility test, for instance, it was possible for a central agency to judge the value of training in voice communications at all communications schools by quantitative methods. Achievement and proficiency tests made it possible to centralize examination procedures, to spot good and poor schools, and to standardize school output.

The same kind of procedures can be applied to operational training in fleet units of the Navy or field units of the Army. This general type of procedure has long been used in Navy gunnery and engineering exercises for ships or crews as a whole, but it has seldom seen field use for individuals and it has never been used in the field with the refinement which is now possible. Achievement and proficiency testing should be extended for use in operational training in as many service units and duties as possible.

Achievement and proficiency tests should also be used more widely in advanced classification. Because Army and Navy recruits are young, the initial classification and assignment of men on induction must be based chiefly on aptitude. At the end of any tour of duty, reassignment occurs. At present, reassignment is based chiefly on initial assignment, corrected only by a rather unsystematic appraisal of performance on duty. Achievement and proficiency tests could provide a basis for systematic reassignment based on objective information as to the gain or loss in knowledge and skill during the duty period. In specific instances connected with radio code, radar, and lead-computing gunsights, the Panel developed proficiency tests for the purpose of advanced classification. The systematic development of this field is indicated.

CHAPTER 9

THE FUTURE OF MILITARY PSYCHOLOGY

THE next war, the pessimists say, will be a pushbutton war. In this catch-sentence they suggest that the next war will be impersonal, that men will be replaced by machines, that the men with the biggest and fastest rockets or the most devastating atomic bombs will surely win the next war. There is an awful truth in this suggestion, but it is a half-truth. The concept of a pushbutton war ignores these facts: the next war will be started by men in order to impose their wills on other men; it will be preceded and accompanied by large-scale psychological warfare; and it will be won or lost by the true fighting unit, the man-machine. In our fascination with the physical sciences and industrial technology let us not forget that the machine is only half the battle. No one has yet invented the pushbutton man, and military psychology is not intended to produce such a man. But there is a science and technology of man as well as a science and technology of the machine.

Wars are won and lost neither by men nor by machines alone. Failure of either component produces failure of the other and a lack of harmonious adaptation of each to the other will lose the next war. The military psychologist should be the scientific bridge between the two widely different elements, equipment and personnel. In World War II the Army, the Navy, and the scientific world realized this too late for maximal effectiveness. As a result probably the most important contribution of the psychologists of World War II was to demonstrate that the man-machine, rather than the machine alone, is the fundamental fighting unit. If we are to maintain our military strength against possible enemies, the lesson of World War II must not be forgotten. Military psychology must continue to grow from the strong beginnings already made. What, then, is the future of military psychology?

This was the question that the psychologists of the Applied Psychology Panel asked themselves as World War II drew to a close and preparations were made to return to civilian pursuits. It was thought that the best way to assure the future of military psychology would be to examine the past accomplishments of the Panel to convince the Army and Navy that such work should be continued in the future. Accordingly the Panel arranged[1] a Joint Army-Navy-OSRD Conference on Psychological Problems in Military Training, which summarized the contributions of psychology to military proficiency, including not only the Panel contributions but also the contributions of other military psychologists.

The Conference held four sessions devoted to the following subjects: I. Simplifying the Task of Military Training (the design and operation of equipment); II. How to Train; III. Who Should Be Trained; IV. Measuring the Effects of Military Training (achievement and proficiency testing).[2] The Panel described its own techniques which had provided the following knowledge:

1. How to save many thousands of man-years in the preparation of men for combat.
2. How to increase combat efficiency by a large factor. (The size of the factor cannot be estimated but it is at least as great as that resulting from many expensive improvements in materiel.)

[1] The Panel was assisted in organizing the Conference by the War Department Liaison Officer for NDRC and the Navy's Office of Research and Inventions.

[2] The Proceedings of Part I of the Conference, *Simplifying the Task of Military Training*, were reproduced as OSRD Report No. 6079 and distributed by the War Department to many of its manufacturers. This report is not available to the general public. The Proceedings of the Conference as a whole were not reproduced for publication despite the fact that with a few exceptions they were released from military security in 1946. The released portions of the Proceedings are held intact in the files of the Applied Psychology Panel, NDRC.

a. How to improve the coordination of different units through reduction in delay and misinterpretation of communications.
b. How to drop more bombs within a target area.
c. How to increase the percentage of field artillery shells falling squarely on enemy targets and how to decrease the percentage falling on our own troops.
d. How to detect enemy planes at greater distances and follow them more accurately during their attacks.
e. How to shoot down more enemy planes, whether by ground, ship, or airborne guns.
f. How to improve the performance of many specific kinds of personnel, including:

> Gunners and firecontrolmen
> Firemen and engineers
> Radio operators
> Radar operators
> Amphibious crews
> Lookouts
> Torpedomen
> Quartermasters
> Signalmen
> Electricians

3. Of greatest importance, how to measure military proficiency so that still further improvements in the man-machine combination may be developed in the future on a basis of sound experimentation.

In summarizing the conference two service officers discussed the results and referred to the future. Rear Adm. W. S. Delany, Assistant Chief of Staff (Readiness), Headquarters, Commander in Chief, U.S. Fleet, and Maj. Gen. I. H. Edwards, War Department General Staff, G–3, addressed the meeting. Admiral Delany said, in part:

"May I first take the opportunity to thank you on behalf

of the Navy for what psychologists individually and as a group have done to help the Navy with its personnel problems during this war. The contributions that you have made to our problems of selection, classification, and training have been invaluable. From the cross section of American youth you have made square pegs for square holes and round pegs for round holes; and you have helped us to fit them together into a completely satisfactory fighting and working organization. You helped us by prescribing standards for selection of men and in the development of various training devices. Without such assistance we never could have put so many men in the right jobs in the fleets, nor put such well-trained crews in their entirety on so many brand new ships. . . .

"I trust that the Navy may have the advantage of your ideas in matters of this sort in the years to come. It is my sincere hope that some of you will be willing to advise with the services from time to time in years ahead. I refer not only to the civilians but also to those psychologists who have so capably served as reserve officers during the war period."

And General Edwards said, in part:

"Such a conference as you have held these past two days has come at a significant time for taking stock of training accomplishments and forecasting training developments. We have won smashing victories over enemies who were highly trained, and who had a discouraging head start over us. There is reason to evaluate, while it is fresh in mind, what we have already done. But we can look forward as well as back, because the task of military training is far from being ended by the technological developments recently disclosed. One thing this war has proved conclusively: however well equipped an army is with the mechanical marvels of today, it cannot win without well-trained manpower. As a matter of fact, the more intricate the machinery, the more highly trained must those be who operate it and keep it in instantaneous working order. Assuming, then, that so long as there are wars or the possibility of wars, men must be militarily trained, we have a large peacetime job ahead of us, even if we have virtually completed our immediate training assignment.

"Following the outline of your conference, let me make a few brief remarks on what seems to me most significant as regards training problems in the army. First, the relation of the design of military equipment to the training of the soldiers who must operate and maintain such equipment. It is quite conceivable that modern science could produce weapons which, despite their marvelous offensive or defensive capabilities, we could not use in numbers against an enemy. They might be so complicated that we would be unable to train sufficient operators, in the time at our disposal, to man them. . . . The enemy does not wait for us to get ready. Extra weeks and months required in training may lose battles and even wars. This will be even more true in the future, when the tempo is certain to be greatly increased. So, from the training standpoint, simplification and consideration of the individual in the design of equipment, are of real concern.

"The training task that has been performed by the Army during the war years has been tremendous, however you measure it: by number of students, number of instructors, or scope of subjects. In this huge enterprise, and with speed of preparation for military duties paramount, it would have been easy to fall into a routine mechanical production of soldiers. The individual might well have been lost in the mass. But it would not have given us the quality of product we needed. With the differences in individuals and with the differences in specialties, with the need also to discover leaders and instructors, we have found it well to use many methods of training rather than one. We have analyzed the individual and analyzed the job, and tried to fit the two together. . . .

"Several of you have discussed the problem of measuring the effectiveness of training. The importance of such testing is twofold: to determine whether the method of training is giving the best results or should be revised, and to find out whether those who have gone through a certain course are.

ready to perform the duties for which they were trained. When it is considered that men are being made ready to engage in the life-and-death business of war, the full significance of this is seen. It must be very sure. . . .

"In conclusion, let me say that it is reassuring to know that we shall have the help of psychological research in solving the military training problems of the postwar period. These problems will probably be somewhat different from those we have encountered in wartime. But I venture to say that, in their way, they will be no less challenging. You psychologists will have a continuing contribution to make."

The words of these high officers summarized the discussions of many of the Panel's liaison officers during the Conference. Representatives of the Army, the Army Air Forces, and the Navy repeatedly suggested that military psychology should continue to develop and make its contributions in the postwar period. Nevertheless many psychologists expect that military psychology will die away in the days to come, as it died away in the days following World War I. Military psychology will certainly die away unless the services themselves take active steps to preserve it. A relatively passive desire for its preservation will not be enough. The future of military psychology rests primarily with the Army and Navy.

In the years to come the Army and Navy may have practically any kind of research program in the field of military psychology that they desire. Budgetary considerations are of minor significance in this respect. The yearly cost of a reasonable psychological program in any one of the services should be less than the cost of a single very heavy bombing plane lost through pilot error. In the work of the Applied Psychology Panel the yearly cost of a single research psychologist was less than the cost to the Navy of a single student pilot who washed out at the end of training. A psychological research program may be expected to pay for itself even if we do not count the remote values which should accrue in a hypothetical future war. In consequence the Army and Navy are free to undertake whatever program of reasonable size that they desire. The possibilities are outlined below.[8]

Prior to World War II it was the frequent experience of psychologists interested in military problems to be told that the Army and Navy welcomed the opportunity to consult them but needed no research assistance. This attitude persisted throughout the war years in the minds of some officers, but it was by no means universal, for research was not only welcomed but eagerly sought in a number of commands and on a number of problems.

Both the Army and the Navy know a great deal about personnel; both have a long history of experience and a record of success. In addition, both have the benefit of recent advice from expert consultants and the results of wartime research studies to aid in handling personnel during peacetime. It might therefore be decided, either as a matter of general policy or for a particular field, that no further research is necessary.

The Army and Navy frequently employ expert consultants to advise on special problems. On a number of occasions the Applied Psychology Panel was asked to furnish a man well acquainted with a particular type of military problem who could serve as an expert consultant for a few days' time. During these short periods of consultation the man was expected to survey, for instance, a training installation, observe the instruction, and recommend improvements in instructional material, course organization, lesson plans, and examinations.

The results of such consultations varied, depending upon the experience of the consultant and upon the similarity between the problems involved in the military situation and those with which he had had previous experience. The best of men may

[8] The following discussion is summarized from two sources: *STR* I, Foreword by C. W. Bray, for which special acknowledgment should be made of the assistance of Dael Wolfle; and C. W. Bray, Psychological Research in the Army Air Forces, pp. 117–137, in W. R. Lovelace, II, A. P. Gagge, and C. W. Bray, *Aviation Medicine and Psychology: A Report Prepared for the AAF Scientific Advisory Group.* Dayton, Ohio, Headquarters, Air Materiel Command. May 1946.

come into a service situation with no direct experience, or experience only from the remote past, of the details of service operations. He can deal only in generalities and the value of his recommendations will depend on the psychological skill and understanding of those who employ him. In contrast, specific and detailed recommendations can be made by an expert who has had the benefit of months of research work in a military training or operating area. The recommendations of such an expert can often be implemented without further research.

Consultation on special problems is the second type of opportunity available to the Army and Navy.

As a third possibility, psychologists can be used to maintain the established personnel objectives and procedures. Psychological procedures, like others, require continuing attention or they gradually become ineffective. Failure to maintain personnel methods results in a loss of the values originally attained.

A program of maintenance is necessary, but if activity is limited to maintenance alone, any mistake in original plans may persist unchallenged and new developments may pass unheeded.

The final type of program which may be considered consists of a thoroughgoing continuing research analysis of the interrelated problems of military psychology. These problems are the problems of classification, training, and equipment, together with the associated problems arising in the clinical and social psychology of service personnel. Cooperative and coordinated research on all the problems of military psychology is the most efficient type of program and may normally be expected to lead to the greatest improvements in military proficiency.

If the Army and Navy desire a complete research program they should give due consideration to wartime experience. In the case of the Applied Psychology Panel it was the opinion of the Panel staff that the project on lead-computing gun-

sights (see Chapter 6) might well be considered as a prototype for the organization of future research. The success of
that project is worth analysis. The success revealed at least
the following:

1. The existence of a real need for psychological study of
   new equipment.
2. That psychological research must be carefully organized
   under sympathetic supervision and established in the
   most effective time and place.
3. The caliber of assistance and the amount of time required from regular service officers. This included assistance in securing research facilities, diversion to the
   project of at least one preproduction model of each new
   device, participation in the work itself, and, particularly,
   active steps to insure application of results.
4. The need for education in technical subjects and the
   psychological training of the project staff.
5. That the research group should have prior experience
   in military research.

These characteristics of the organization of the project
should be seriously considered in any postwar program. Unless there is a real need; unless the research is done at the
right time and place; unless the services are prepared to try
to understand, to give active assistance, and to trouble themselves to apply results; and unless the research personnel
and their representatives at higher echelons are professionally
capable and experienced in military conditions, much is
wasted that might be gained from a research program.

If the Army and Navy desire a complete program they
must secure men with adequate training in experimental psychology. Competent psychologists can be obtained either by
establishing positions in the services or by contracting for
research in civilian units such as the universities.

Research under contract can secure the help of capable
psychologists. Their value to the services should be great if
they are familiar at first hand with service problems, but it

is doubtful that part-time university-based workers can maintain the degree of understanding of service conditions required to make important practical contributions to military psychology. The practical success of contract research in military psychology will depend on the guidance provided by the services, but guidance is likely to be effective only in proportion to the psychological understanding of those who guide. The services must maintain their own staffs of capable psychologists if they desire a complete research program.

Whether or not capable psychologists will accept service positions depends upon the conditions offered for experimental work. Capable psychologists will not enter the Army or Navy during a peace unless there is good assurance that they will be able to work effectively on research problems. That assurance depends upon freedom from the danger of being sidetracked into routine administrative duty. It depends upon freedom from casual interference by those who are not technically trained in psychology. It depends upon long-term planning; year-by-year commitments are acceptable during war, but not in planning peace-time research. It depends upon the ability of the military psychologist to maintain his status as a recognized member of the profession of psychology.

Whether or not capable psychologists will accept service positions in peacetime depends, then, upon the organization of military psychology within the services. Only a general psychological research agency within the services or within a major branch of the services can surely provide the conditions specified above. Only a general research agency can provide a continuing high staff level of military supervision so that direction will be in the hands of officers with an appreciation of the broad implications of psychological research and with the authority to see that suitable research facilities are available. Only a general research agency can provide freedom from diversion of psychologists into non-psychological work. If psychologists are scattered through the branches of the services; if they are attached, a few at a time, to a variety of administrative, scientific, engineering, or medical units; and if they are not represented in higher echelons by psychologists, they tend to become non-psychologists. As a result much of their specific value is lost. This is not to say that there should not be close and continuous co-operation between psychology and the other disciplines by means of inter-disciplinary groups. Nor should it prevent other scientists from hiring occasional psychologists as technicians, just as psychologists hire engineers, mathematicians, and physiologists as technicians. But psychology is a science in its own right; it is a young science and it may best continue to develop and to assist the services if it is left primarily in the hands of those who know it best.

Since the end of World War II several hundred permanent positions have been created in the services for military psychologists. At the time of writing, only a handful of the thousand or more psychologists who were engaged in wartime psychological work have accepted these permanent positions. Salaries, frequently fifty per cent higher than the universities offer, are not attractive because of the belief that the conditions and organization described above either do not exist or, if they exist, that they will not endure. Of the few who accepted service positions at the end of the war, a number have since returned to university work.

The Army and Navy can have any type of assistance they desire. Whether either service will have an effective research program in military psychology depends upon the availability of personnel with adequate training in experimental psychology. Competent psychologists will be available when satisfactory conditions for experimental work are known to exist.

Conclusion

The interest of the armed services in military psychology should be continuous. Service recognition of this fact is illustrated by the words of a battle-scarred and beribboned

Admiral present at a demonstration of the mechanical and electronic marvels of the U.S.S. *Missouri* to a group of distinguished engineers and scientists. His words, as he swept his hand over the deck of the pride of the Fleet: "Twenty-five hundred officers and men. Gentlemen, twenty-five hundred sources of error."

It was the purpose of the Applied Psychology Panel and of many other psychological war agencies to help the Army and Navy reduce human error during World War II. The success of future psychological efforts to reduce human error in the military situation—the success of the future development of military psychology—rests with the Army and Navy.